For Barbara

This book was made possible by NASA Living With a Star grant number NNG06EC631

The graphic image of the Sun on the cover and in the text is a detail from a much larger woodcut created by Albrecht Dürer (1471-1528) to illustrate the prophecies of the Apocalypse in an early book printed in Germany in 1498.

NP-2009-1-066-GSFC

For sale by the Superintendent of Documents, U.S. Government Printing Office
Internet: bookstore.gpo.gov Phone: toll free (866) 512-1800; DC area (202) 512-1800
Fax: (202) 512-2104 Mail: Stop IDCC, Washington, DC 20402-0001

ISBN 978-0-16-083807-1

Dear Reader,

The enclosed work represents a loving and dedicated effort on the part of my beloved late husband, John A. Eddy, known to most of us as Jack. With the exception of a few minor administrative details, the book was only just completed, and the final, final version signed off on, when our three-year battle with cancer took him from us on June 10th 2009. And although Jack, sadly, did not get to see the final printed version, all of the words, illustrations, tables, graphs, covers, even the most minor details, were carefully and thoughtfully conceived, written, and edited by him over a multi-year period in our peaceful home office in Tucson, Arizona.

It was Jack's hope and goal to assemble in one place, through this book, concise explanations and descriptions—easily read and readily understood—of what we now know of the chain of events and processes that connect the Sun to the Earth, with especial emphasis on space weather and Sun-Climate. As many know, Jack was extraordinarily gifted at teaching and illustrating and popularizing any subject, but particularly those in the realm of science. I believe he once again shared these gifts with us through his many years of work on this book.

For those of us who were blessed to know Jack, we will surely always remember his kind, gentle, soft-spoken and dignified nature as well as his magical sense of humor. And all will sense these traits within the thoughts and words that he has given us through this book. I am deeply proud and honored to have been his indirect and unspoken partner in this endeavor, as well as his wife, soul-mate, and life partner for nearly twenty years.

Barbara
_____
Mrs. John A. Eddy
Tucson, Arizona
July 4th 2009

The Sun-Earth System: CONTENTS

AN OVERVIEW

The Stars Around Us	1
Our Dependence on the Sun	3
The Sun's Inconstancy	3
Intruders from Afar	5
What Gets By	5
Voyages of Discovery in an Age of Exploration	6
A New Appreciation	7
The Consequences	8
An Interconnected System	9

THE SUN

The Sun as a Star	13
A Voyage to the Sun	13
Perpetual Combustion	14
The Hidden Source of Solar Energy	16
Delayed Delivery	16
Radiant Energy from the Sun	17
How Constant Is and Was the Sun?	18
Metered Sunshine	19
The First Who Saw the Face of the Sun	20
The Long Watch	23
The Sun That We Can See	24
The Photosphere	24
Sunspots	26
Bright Faculae	28
Beneath the Shining Surface: The Bubble Machine	30
Lifting the Veil: The Unseen Sun	31
The Sun's Chromosphere and Corona	33
How We See the Corona and Chromosphere	38

THE SOLAR WIND & SOLAR VARIABILITY

The Solar Wind	45
Sources and Characteristics of the Solar Wind	47
Solar Variability	50
Why the Sun Varies	52
Short- and Long-Term Changes in Solar Activity	56
Solar Explosions and Eruptions	57
Explosive Solar Flares	57

Solar Prominences and Filaments 60
Coronal Mass Ejections 64

THE NEAR-EARTH ENVIRONMENT

A Protected Planet 71
The Air Above Us 72
Changes on the Way to the Top 72
The Troposphere 75
The Stratosphere 76
The Mesosphere and Thermosphere 77
The Ionized Upper Atmosphere 79
The End of the Atmosphere 81
Into the Magnetosphere 82
The Form and Function of the Magnetosphere 84
The Paths that Particles Follow 85
Captive Particles in the Magnetosphere 86
The Earth's Radiation Belts 88
The Plasmasphere 91
The Heliosphere 92
Cruising the Heliosphere 94

FLUCTUATIONS IN SOLAR RADIATION AT THE EARTH

Changes in Total Solar Irradiance 99
Variability in Different Parts of the Spectrum 102
Effects of the Sun's Rotation 104
Effects of the Earth's Orbit 105
Lost in Transit: The Fate of Solar Radiation in the Earth's
    Atmosphere 107

VARIATION IN THE FLOW OF PARTICLES AT THE EARTH

The Nature of Arriving Particles 111
Solar Sources 113
Particles Borne Outward in CMEs 113
Particles from Solar Flare 115
The Solar Wind Plasma 117
Characteristics of Slow Solar Wind Streams 118
High-Speed Solar Wind Streams 118
Sectors in the Sun's Extended Magnetic Field 119
Pushing and Shoving on the Way to the Earth 121
When Solar Particles Strike the Earth 122

Through the Guarded Gates ... 124
Magnetic Reconnection ... 126
Effects of Changes in the Earth's Magnetic Field ... 129
Cosmic Rays ... 131
The Fate of Cosmic Rays ... 132

IMPACTS OF SOLAR VARIABILITY

Solar Causes, Terrestrial Impacts, and Societal Effects ... 139
Impacts on Near-Earth Space ... 141
Magnetic Storms ... 143
The Aurora ... 146
Impacts on the Upper Atmosphere ... 153
Perturbing the Earth's Electric Field ... 155
Restructuring the Ionosphere ... 156
Disturbing the Biosphere: The Lower Atmosphere, Oceans,
    and Land Surface ... 159

EFFECTS ON HUMAN LIFE AND ENDEAVOR

What is Affected ... 165
Some Specific Societal Effects ... 165
Exposure of Aircraft Passengers and Crews ... 167
Risks to Manned Space Flight ... 171
The Ocean of Air ... 171
Enhanced Ultraviolet And X-Ray Radiation ... 172
Solar X-rays ... 172
A Sun Intensely Bright ... 172
Solar Energetic Particles and Cosmic Rays ... 173
The Physiological Effects of Ionizing Radiation ... 176
The Importance of Dosage ... 176
The Disaster That Almost Happened ... 178
Impacts on Spacecraft, Space Equipment and on Observations
    of the Earth From Space ... 182
Times of Particular Hazard ... 184
Flight Paths of Greatest Risk ... 184
Spacecraft at the Lagrangian Points of the Sun-Earth System ... 185
Polar Orbits and the South Atlantic Anomaly ... 186
Geosynchronous and Geostationary Orbits ... 186
Destructive Particles From the Sun and the Earth's
    Radiation Belts ... 187
Cosmic Rays ... 188
Atmospheric Drag ... 189
Impacts on Micro-Circuits and Computer Systems ... 191
Damage to Other Space Equipment ... 192

Protecting Against Damage From High-Energy Particles   193
Impacts on Telecommunications, GPS, and Navigation   195
Direct and Indirect Reception of Radio Waves   196
Role of the Sun and Solar Variations   197
Impacts on GPS and Other Navigation Systems   198
Effects on Electric Power Transmission   199
The Power Blackout of 1989   200
How Magnetic Storms Disrupt Power Systems   201
Where Solar-Driven Power Outages Most Often Occur   201
Effects of Geomagnetically-induced Currents on the
   Cost of Electricity   203
Early Signs of Solar Interference in Communications   203
Effects of GICS on Telecommunications Cables   204
Damage to Pipelines   205
Impacts of Geomagnetic Storms on Geological Surveys
   and Explorations   205

EFFECTS OF THE SUN ON WEATHER AND CLIMATE

A Brief History   209
The Missing Pieces   212
Metering the Energy the Earth Receives   213
Recovering the Past History of the Sun   217
Effects of Solar Spectral Radiation   219
Sensitivity of Climate to Solar Fluctuations   221
11-year Solar Forcing   223
Solar Forcing of the Oceans   224
Hidden Diaries of the Ancient Sun   225
The Fate of Carbon-14   227
Beryllium-10 in Ice Cores   228
Marks of the Sun on North Atlantic Climate During
   the Last 11,000 Years   230

FORECASTING SPACE WEATHER AT THE EARTH AND BEYOND

Space Weather   235
Predictions   236
Sources of Needed Data   238
Available Warning Times   240
Especial Needs for Manned Space Exploration   242
Current Capabilities   245
Operational Facilities   246
The Heliophysics System Observatory   249

REFLECTIONS

Solar Misbehavior	255
What Has Changed?	256
The Sun and Global Warming	258

ACKNOWLEDGMENT — 261

APPENDICES

Glossary of Technical Terms	263
Sources for Additional Information	291
Tables	295
Images and Illustrations	297
Index	301

The Great Andromeda Nebula, a vast collection of stars some 2 million light years away. Although each of the stars in the central core and spiral arms of that faraway galaxy is much like our own star, the Sun, we see most of them blurred together by great distance, and through a speckled foreground of countless other, much nearer stars which belong to own spiral-shaped galaxy, the Milky Way.

AN OVERVIEW

The Stars Around Us

In a world of warmth and light and living things we soon forget that we are surrounded by a vast universe that is cold and dark and deadly dangerous, just beyond our door. On a starry night, when we look out into the darkness that lies around us, the view can be misleading in yet another way: for the brightness and sheer number of stars, and their chance groupings into familiar constellations, make them seem much nearer to each other, and to us, than in truth they are. And every one of them— each twinkling, like a diamond in the sky—is a white-hot sun, much like our own.

The nearest stars in our own galaxy—the Milky Way—are more than a million times farther away from us than our star, the Sun. We could make a telephone call to the Moon and expect to wait but a few seconds between pieces of a conversation, or but a few hours in calling any planet in our solar system. But one placed to the nearest star would impose a mandatory wait of almost nine years between a question asked and the answer received. And though the radio waves that carry our phone messages travel at the unimaginable speed of 670 million miles per hour, calls placed to the celestial area codes of most of the other stars in the sky would exact truly big-time roaming charges, with conversational exchanges separated by unavoidable pauses of hundreds to thousands to millions of years.

Yet these vast and lonely reaches of interstellar space are neither vacant nor ever still. There are about 30 atoms—mostly hydrogen and helium—in each cubic foot of interstellar (or interplanetary) space, and many more than this in the vicinity of the Sun and the planets. In addition, there are also occasional molecules and dust grains. Moreover, were we to look closely at any sampled volume, however small and wherever it is, we would find it heavily trafficked: continually crossed and criss-crossed by high-speed, atomic particles from the Sun or the cosmos that are ever passing through.

Many of these itinerant transients are the accelerated protons and electrons and atomic nuclei that are flung outward continually into space from the overheated atmospheres of our own Sun and countless other stars: the omnipresent effluent

Technical or semi-technical words introduced in the text are printed in blue generally where they first appear in the text. They are then defined in alphabetical order in the Glossary, starting on page 263.

of solar and stellar winds. Others—the so-called cosmic rays, which are often more energetic and hence more lethal—are atomic particles of the same kinds that were expelled much more forcefully from catastrophic stellar and galactic explosions of long ago and far away.

Our Dependence on the Sun

The Earth, warmed and illuminated by a benign and nearby star—and insulated and protected from a surrounding environment that is entirely hostile to life—remains a most unusual island in the middle of a dark and stormy sea.

Life itself is possible only because of the Sun and our nearness to this strong and unfailing source of light and heat and energy. From it, and for free, we receive in endless supply not only light and heat, but a steady stream of countless other gifts and essentials. Among them are the blue of the sky, the clouds, rain and snow, trees and flowers and tumbling streams, the replenishment of the oxygen we breathe, and all of the food we eat.

The Sun is the source of all the energy that human beings have ever used in burning wood, and in burning coal, oil, gasoline and natural gas: for these fossil fuels are but repositories of solar energy from ages past, captured through photosynthesis in the leaves of plants from long ago. All the energy we derive from wind and water power comes also from the Sun. For it is heat from the Sun that drives the winds, and heat from the Sun that keeps the rivers flowing by evaporating surface water from the seas, to fall again as rain on inland watersheds.

Can there be any wonder that the Sun was so widely revered and sanctified by early peoples everywhere? Or that hymns still sung in churches today so often draw on solar similes?

> *Break forth o beauteous heavenly Light, and usher in the morning*
> *Come, quickly come, and let Thy glory shine, gilding our*
> *darksome heaven with rays divine*

The Sun's Inconstancy

Yet the Sun is neither constant nor entirely beneficent. As a variable, magnetic star, the Sun is ever changing and in many ways: most often through violent explosions and eruptions of colossal scale. The glowing, gaseous surface that appears so perfectly white and still from far away is in reality a roaring, roiling arena of continual conflict and eruption.

Even the sunbeams that stream outward in such bounty are not entirely benign: for with them come solar gamma rays, x-rays, ultraviolet radiation, and highly energetic atomic particles, all of which are potentially lethal for living things. Were it not for shielding by ozone, atomic oxygen and nitrogen in the upper atmosphere—and the armor of the arching lines of force of the Earth's magnetic field—damaging short-wave radiation and atomic particles from the Sun would have long ago extinguished life on Earth. Sunlight allowed life to take hold on the planet, but it is these invisible shields, high above our heads, that have permitted it to continue and evolve.

Most of the Sun's short-wave radiation is kept from reaching the surface of the Earth and the oceans: absorbed by atoms and molecules of air as it streams downward through the atmosphere. The absorbers of gamma rays and x-rays and far ultraviolet radiation are chiefly atoms and molecules of oxygen and nitrogen that populate the thin and highly rarefied air in the upper atmosphere, as far as 100 miles above the surface.

But the bulk of the Sun's damaging ultraviolet radiation gets through these first lines of defense and penetrates further, until with as little as five miles to go before it reaches the surface, it is finally stopped. There in the stratosphere and upper troposphere, it is largely absorbed and blocked by molecules of ozone: a trace gas present in so meager a supply that were all of it compressed to the density of ordinary air at the surface of the Earth it would be much thinner than a window pane.

The recurring streams of atomic *particles* that flow continually outward in heated winds from the Sun—or in violent bursts when parts of the Sun erupt and explode—are blocked by other means. Incoming atomic particles, regardless of their electrical charge, are fast depleted in number by repeated collisions with atoms and molecules in the air, from the moment they enter the Earth's atmosphere.

Most of those that carry a + or − electric charge—including protons and electrons and ions of helium and lithium and boron and almost all the other elements—are repelled or captured before they can enter the upper atmosphere of the Earth. In this case the shielding—which is most protective at lower latitudes—is provided by the arched lines of magnetic force that tower high above the surface of the planet, and all around it, providing a wrap-around magnetic bumper, rooted in the planet's internal magnetic field. Undeterred by this magnetic barrier are high-energy galactic cosmic rays, whose energies per particle are so great that they pass right through it and on down into the upper atmosphere.

Intruders From Afar

Interstellar particles from a host of other, more distant stars also thread the solar system, as do the considerably more energetic particles that arrive as cosmic rays from unseen stellar and galactic explosions elsewhere in the universe. Charged galactic cosmic rays of sufficient energy per particle can overpower the defenses of our own magnetosphere to spend their prodigious energies in collisions with atoms and molecules of air in the upper atmosphere. Many of these, however, are repulsed by the Sun before they reach the near vicinity of the Earth, as though shooed away by a protective mother hen.

The degree of cosmic ray protection we receive from the Sun depends upon magnetic conditions on the surface of the star, which wax and wane in a cycle of about eleven years. In the peak years of the cycle, when the extended realm of the Sun (the heliosphere) is more disturbed, more alien particles are turned away. In the valley years, at the minima of the solar cycle, there are fewer disturbances in the heliosphere, and more cosmic rays intrude into the solar system.

The last maximum of solar activity was reached in 2001. Today, as solar activity has fallen to an apparent minimum in the 11-year cycle, the traffic of galactic cosmic rays in the near-Earth environment is especially heavy. A few years from now, when solar activity begins another climb—toward an expected sunspot maximum in about 2013—the number of cosmic rays that reach the Earth will again be diminished by about 15 percent.

What Gets By

Neither the magnetosphere nor the atmosphere is 100% successful in blocking all that the Sun (or other more distant sources) sends in our direction. Usable heat and light are allowed to pass, almost unchecked, as we would have them do. As noted below, some of the Sun's ultraviolet radiation, though depleted, also makes its way to the ground.

Some of the charged solar particles that approach the Earth still find their way into the upper atmosphere and magnetosphere. Most are caught and temporarily stored within closed field lines of the Earth's magnetic field, but these, too can work their way, in time, into the upper atmosphere.

When these and other energetic particles collide with atoms of air about 100 miles above the surface, they provoke atoms of oxygen and nitrogen and certain molecules to emit light in pure colors of green or red or blue. When conditions are right, we can witness the intrusion of these escapees from either the Sun or the Earth's magnetospheric holding cell as brightly painted streaks of light in the northern and southern sky: the transitory and ethereal displays of the aurora borealis and aurora australis.

Nor are all the harmful rays of sunlight blocked from reaching the ground. A fraction of the damaging solar ultraviolet radiation that pours down unremittingly on the dayside of the Earth slips by the stratospheric ozone shield, to continue downward as potentially-damaging, shorter-wavelength UV-A and UV-B radiation.

The number of UV-A and UV-B rays that make it to the surface of the Earth depends on the variable thickness and density of the stratospheric ozone layer; on the distance the ultraviolet rays must travel through the remaining air that lies beneath it; and on the clarity of the sky. It depends as well on how much the Sun itself has generated, for solar ultraviolet radiation varies from day to day and year to year, in step with changes on the surface of the Sun. But on any day, a much greater fraction of damaging solar ultraviolet rays will reach the ground (1) at high elevations (as in the Rocky Mountain west); (2) where the Sun most often shines (as in southern Arizona); and (3) at lower latitudes (as in Florida, Hawaii, or Australia) where sunlight passes more vertically through the air and hence along a shorter path to reach the surface.

Voyages of Discovery in an Age of Exploration

The story of how we came to know all that happens just beyond the protective walls of our atmosphere and magnetosphere is a tale of recent exploration. As is that of how the near-Earth environment, and the walls themselves, are shaped and battered by a highly variable Sun. While much had been learned about both the Sun and the Earth, the space between them was, before the second half of the last century, largely *terra incognita*: a void sketched in with theory and supposition, much like the early maps of the known world that were drawn before the epic voyages of Columbus and Magellan.

Although rocket-launched probes had briefly sampled the nearer reaches of the upper atmosphere in the late 1940s, it was about the time when NASA was established, in 1958, that longer and farther voyages of exploration at last began.

The first of these were lofted into low-Earth orbits but several hundred miles above the ground, to circle the planet at the ragged upper edges of our atmosphere. Here only the last vestiges of air remain; the sky is no longer blue; and the atmospheric pressure is reduced to less than a millionth of a millionth of what we are accustomed to at the surface of the planet.

Others ventured further, into and beyond the Earth's extended magnetosphere, charting for the first time its form and content and asymmetry. Immediately encountered in these early explorations were the Earth's radiation belts: largely unexpected swarms of captured electrons and protons and other charged particles, entrapped by the Earth's magnetic field and held there, as in a magnetic cage.

The shape of the Earth's magnetic field was found to be not the simple curved lines of force like those traced out by iron filings above the poles of a bar magnet, but a dynamic, distorted and distended structure: compressed flat on the Sun-facing side by the persistent pressure of the solar wind, and stretched out by the solar wind on the other side to well beyond the orbit of the Moon.

Further explorations into near-Earth space identified, at last, the solar sources and ensuing chain of events responsible for the disturbances, called geomagnetic storms, that had been detected and recorded by magnetic instrumentation on the surface of the Earth since the middle 1800s.

Polar aurorae—the occasional displays of colored light in the nighttime sky that had been described in lore and legend for at least 2000 years—were seen in their entirety from the vantage point of space. Observed from above, what had appeared from below as curtains of colored light were revealed in their entirety as full rings of varying diameter, centered on the Earth's magnetic poles.

There, much like a pair of neon signs, glowing circles of light define the zones near the two poles of the planet where electrons and protons most easily find their way down into the Earth's upper atmosphere. And there—hundreds of miles high—they spend their energies in aerial collisions with atoms of oxygen and nitrogen. Some of these energetic particles come directly into the atmosphere from the Sun; others from a cache of stored particles—some captured earlier from the Sun, others from our own ionosphere—that are sequestered in the Earth's magnetotail.

A New Appreciation

In these voyages of discovery much was learned of the true nature of the Sun, the real character of the near-Earth environment, and the form and functions of the magnetosphere and the upper and middle atmosphere. But the truly New World that was found in all these explorations was surely that of the winds of charged particles that blow outward without respite from the Sun, in all directions, including our own. At times these blow but briskly. At others, they come far faster, driven by Krakatoan eruptions on the Sun that tear whole parts of the star away and send them hurtling through the realm of the planets.

The principal legacies of this new age of discovery are a better definition and more enlightened awareness of the very real effects of solar variations on the planet on which we dwell: the realities that accompany the benefits of living with a star. We suspected that the Sun's total radiation varied: we now know it does and by how much. And we have found that near-Earth space—just beyond our door—is a place of tempestuous and often violent weather: not the familiar weather of wind and rain and sleet and snow, but the space weather of variable solar radiation, streams of energetic atomic particles and imbedded magnetic fields, driven by the volatile moods of a near-by star.

The Consequences

Among the down-to-Earth consequences of an inconstant Sun are the impacts of known or suspected variations in the Sun's output of radiation and particles on the climate of the Earth, and the effects of these changes in enhancing or ameliorating the present global heating which is driven principally by increasing greenhouse gases.

As important—societally, economically, and in terms of national security—are the impacts of solar storms and eruptions on communications of all kinds, from telephone to television; on navigational and other geographic positioning systems; on the health and safety of passengers and crews in high-altitude jet-aircraft flights that cross polar and sub-polar regions; on the operation and integrity of electric power grids; on electronic devices of all kinds carried on civil and military spacecraft; and on the safety of astronauts.

What is more and more apparent is that with technological advances and greater sophistication in much of what we do, we lean more and more heavily on systems that are exposed and vulnerable to changes on the Sun and in the near environment of the Earth. Particularly vulnerable are manned space flights, and most particularly those that venture beyond the protective shields of the Earth's atmosphere and magnetosphere; as on envisioned voyages to distant Mars, on shorter travels to and from the Moon, and in connection with the inhabited lunar colony that is now in early planning stages.

Truly essential in all of these impacted areas of everyday life and space exploration is the ability to forecast space weather, day by day and in advance, tailored to specific needs. The accuracy and utility of such forecasts lean heavily on two essentials: (1) a thorough knowledge of the Sun-Earth system; and (2) the availability of an ongoing stream of round-the-clock space weather data.

To meet this challenge, some twenty-six spacecraft are today in orbit around the Earth or the Sun—or on further voyages of discovery, far from home—to explore, patrol and monitor the complex, coupled Sun-Earth system. Their purpose is to track, understand and ultimately predict the major changes on the Sun and in near-Earth space that affect space weather and human endeavor. Together, they make up an ongoing fleet of modern spacecraft, designed and operated to complement each other, and to work together as an ongoing System Observatory in near-Earth space.

An Interconnected System

Our knowledge of the Sun-Earth System has been acquired through the years on a piece-by-piece basis, through the efforts of generations of men and women in many countries. But in these early efforts, the focus was most often on the exploration and study of isolated parts of the larger whole: the Sun, the Earth's magnetosphere, ionosphere or upper atmosphere, and ultimately, with the Age of Space, the nature and dynamics of the interplanetary medium that fills the void between the Earth and its distant Sun.

With new understanding came specialization and ultimately the emergence of whole new fields of study including solar physics, aeronomy, atmospheric physics and chemistry, and space, cosmic ray, magnetospheric, and ionospheric physics. But as was always known, these are all connected together, like links in a chain.

Advances made in recent years regarding the essential interdependence of these elements of the chain, and the pressing need to forecast the impacts on the modern world of solar-driven changes in this system have called for a more holistic approach. And the emergence of the all-encompassing science of heliophysics: the study of an interconnected system, extending from the soil beneath our feet to the white-hot surface of the Sun, and including all that lies between. The ultimate goal is an analytical working model of the coupled Sun-Earth system: a chain of interacting links that can replicate the impacts of solar events on the Earth and human endeavor.

The intrinsic interconnectedness of almost everything, including the Earth and all things in it, is a common theme today, though not at all a new one. "Pull up any part of Nature," said John Muir in the early 1900s, "and you will find its roots entangled with all the rest."

The same truth may have been expressed most eloquently a few years earlier by the mystic poet Francis Thompson, in London, in 1893:

> *All things by immortal power, near or far,*
> *hiddenly, to each other linkéd are;*
> *that thou canst not stir a flower*
> *without troubling of a star.*

We know as well, today, that what happens on our own star—93 million miles away—can perturb the Earth and trouble our own lives and well-being.

The disk of the Sun seen in the EUV in the light of singly-ionized helium, revealing the hotter (about 100,000° F) and more disturbed upper chromosphere that lies above the more placid 10,000° F photosphere which appears to us as a featureless white disk in the sky. The large loop of incandescent gas is a magnetically-formed extension of the chromosphere that reaches far into the higher and ever hotter corona.

THE SUN

The Sun as a Star

The star we call the Sun is to us, of course, unique. Yet it is much like countless others in the sky that burn as bright and for as long. Some stars like Betelgeuse, a red supergiant star in Orion, are more than 500 times larger than ours, and others are a hundred times smaller, no larger than the Earth. Some are older, and many are younger. Among other stars in our Galaxy, our Sun is somewhat atypical in the sense that it is hotter and more massive than 80 to 90% of the rest. Still, were the Sun a person it would be seen by others around it in the sky as large, late middle-aged, reasonably well-behaved, and moderately bright.

The diameter of the Sun is about 100 times that of the Earth: large enough to hold about 100^3, or a million planets the size of ours, were they somehow stuffed inside it. While it doesn't seem that distant, the Sun is 93 million miles away, almost 400 times farther than our nearest neighbor, the Moon.

A Voyage to the Sun

Were we able to fly to the Sun in a commercial airliner at the speeds we are now accustomed to, the one-way trip would take more than twenty years. And after our arrival, should we choose to take a tour, a flight around the solar equator at the same speed would consume another eight months' time.

Once there we would find the star—though unbearably hot and blindingly bright—diffuse and ephemeral: gaseous throughout, like the atmosphere of the Earth, with no real boundary or surfaces. And nowhere to land. All we would see, up close, would be the luminous radiating surface, or photosphere: as bright and hot as a welder's torch, and fully as hazardous to look upon.

As is the case when viewed through a simple telescope from the Earth, the dimmer chromosphere, the graceful loops of solar prominences and the awesome corona, which loom high above the photosphere, would be wholly invisible to us in the awful glare of the searing white-hot, radiant surface. And this life-giving layer in the gaseous Sun—the fount from which all sunbeams flow—would appear not smooth and white but roiled and raging and turbulent: composed of an irregular mosaic of the close-packed tops of gargantuan convection cells, each carrying heat upward from deep within the Sun, like bubbles rising in a witch's cauldron.

Scattered among them, from place to place, we would soon recognize—in the form of larger, slightly dimmer, gaping depressions—the fields of war where magnetic fields, emerging from below, hold back and reduce for a time, though not for long, the inexorable upward flow of convected heat. These are the well-known and long-counted sunspots—many larger than the Earth—whose numbers wax and wane in the course of an inexact cycle of about eleven years.

But nothing we see on the Sun, from near or far, is there to stay: like dancing flames in a fireplace, all that is visible to us is impalpable, ethereal and ever changing. The Sun is something of a paradox in this regard: the oldest, largest and heaviest elephant in the solar system, tipping the scales at more than 1015 trillion tons, and yet—when seen up close—displaying an almost unbearable lightness of being.

What is more, nothing that we see—not sunspots, not the bright surface nor all that lies above it—is as dense as the last vestiges of air that remain at the altitudes where satellites circle the Earth, hundreds of miles high, where the barometric pressure is like that of a vacuum. Truth be told, had we taken our imagined airplane voyage to the star, we could have kept on flying, on the same course that took us there, directly into and inside the Sun. And although the solar interior would get ever denser and hotter the deeper we go, we could fly halfway toward the very center of the star—for eighteen days and another quarter million miles—before reaching anything half as dense as the air we breathe at home.

We could never do this, of course, because of the crushing pressure that continues to increase the deeper we go inside the Sun; and in addition, the searing heat we would have felt long before our solar aircraft ever reached the photosphere.

Perpetual Combustion

What should amaze us is how the Sun can burn so brightly and continuously for so long a time. It is hard to imagine a fire so bright that we could feel its heat a mile away. And harder still the heat from a fire a hundred or a thousand miles away, no matter how large or hot it was.

Radiant heat from any source, large or small—the burner on a stove or the Sun in the sky—rapidly diminishes, by the square of the distance, as one moves away from it. Yet the heat emitted from the Sun—93 million miles away—is so intense that out of doors on a summer day we seek the shade.

The Sun has been shining in this way for about 4.6 billion years, about one third the age of the Universe itself, which is now thought to be 14 billion years. And the Earth—which is almost as old—has for at least the last 3.8 billion years been bathed in very nearly the same level of solar radiation that streams down on the planet today. The amount of solar light and heat that is intercepted by the Earth is about 5500 kilowatts per acre, or almost 2×10^{14} kilowatts over the entire daylit hemisphere.

Because our planet is so small a target at so great a distance, the portion of the Sun's emitted energy that we receive on Earth is a truly negligible fraction of what the profligate Sun pours out in all directions. The total radiative power released by the Sun, day in and day out, is about 4×10^{23} or 400,000,000,000,000,000,000,000 kilowatts. And like every other star in the sky, almost all of it is thoroughly wasted: thrown away and lost forever in the cold and dark of empty space.

THE EIGHT PLANETS AND PLUTO AS SEEN FROM THE SUN

PLANET	DIAMETER IN MILES	DISTANCE RELATIVE TO THAT OF THE EARTH	APPARENT SIZE RELATIVE TO THAT OF THE EARTH
Mercury	3000	.4	1
Venus	7500	.7	1.5
EARTH	**7900**	**1**	**1**
Mars	4200	1.5	0.4
Jupiter	88,000	5	2
Saturn	74,500	10	1
Uranus	31,600	19	0.2
Neptune	30,200	30	0.1
Pluto	1900	39	0.0001

Nor is what we or any other planet receives fully utilized. About 60% of the solar energy that arrives at the top of the atmosphere will make it to the surface of the Earth: the rest is absorbed and put to work in the air, or reflected and returned, unused, back into space. The amount that reaches the land or water at any place depends of course on its latitude and altitude, the time of year, and the clarity of the sky.

For the continental United States, the *average* daily radiation from the Sun that falls on one acre of land is equivalent to the energy released in burning 11 barrels of oil; or in one year, about 4000 barrels.

Were the Sun to bill us at the oil-equivalent rate for the solar energy we receive, the *average* annual *Sunshine Tax*, figured at $100 a barrel, would be $360,000 for each acre we owned; and in some places in the sunny South and West—which receive almost twice as much sunlight as the continental average—well over half a million dollars per acre.

The Hidden Source of Solar Energy

The Sun is indeed huge, but no combustible fuel it might contain could stoke so hot a furnace so long a time; nor could a gradual gravitational shrinkage of the star, which until about a hundred years ago was thought to be the source of solar energy. It was then shown that given irrefutable evidence from geology of an Earth extremely old, the rate of contraction that was needed to supply the prodigious energy flowing outward from the Sun implied a grossly inflated ancient star. In fact, one with so great a girth that at the time when early forms of life were evolving on the Earth, our planet was tracing out its orbital path far inside the star it circled.

The real secret of the Sun's seemingly boundless energy—like that of all the other stars—is instead the nuclear processes that are triggered by the staggering pressures and temperatures deep within its central core: the fusion of single atoms of hydrogen into helium, with the release of some of the energy in each contributing atom in the form of heat. Because the Sun is made almost entirely of hydrogen, which is used up only slowly, this simplest of nuclear processes should continue to provide the Earth with adequate radiative energy for at least another five billion years. At about that time—for those who worry about such things—it is thought that the Sun will have used up its store of hydrogen fuel, and will expand about 100 times in diameter to join the ranks of the so-called red giant stars, like Arcturus and Capella.

Delayed Delivery

The heat created in the nuclear furnace, deep within the core of the Sun, works its way gradually outward to stoke the glowing surface of the star, almost half a million miles above the core. The path to freedom is at first exceedingly slow and tortuous, for the energy released in nuclear fusion is transmitted outward by individual collisions between single atoms of hydrogen and helium, in a medium that is incredibly compressed. To the Sun—that must ultimately expel every erg of newly released heat—the excruciating process of getting rid of it must seem as frustrating as a game of pool, played on a table infinitely long that is tightly packed from bumper to bumper with billiard balls.

The density of matter within the Sun decreases steadily from the core to the surface of the star, and with fewer and fewer collisions farther and farther apart, the initially slow process of heat transfer gradually accelerates with distance from the center. Through the final third of the distance, the energy is no longer passed upward to the glowing surface by collisions between neighboring atoms, but carried up in bulk, through the upward movement of heated matter in giant convective cells. But despite the added speed of this final dash to the finish line, the heat that has finally worked its way outward from the core of the Sun to the visible surface of the star will have spent, on average, about 100,000 years en route.

Once there—free at last—it escapes the Sun as radiant heat: to race onward and outward at the speed of light. In but eight minutes it will reach the orbit of the Earth. There in the warm sunshine of a summer's day, we feel the heat the Sun produced a tenth of a million years ago, when woolly mammoths shared a simpler world with Stone Age men and women.

Radiant Energy from the Sun

Although the Sun also continually showers the Earth with energetic atomic particles, all but about 0.00001% of the energy we receive from the star comes to us in the form of radiant energy spanning a wide range, or spectrum, of wavelengths. About 80% of it is given off in the form of light (the visible spectrum, containing all the familiar colors of the rainbow) and as heat (the infrared spectrum, which we feel but cannot see). Most of the rest comes in a broad span of the more energetic and potentially harmful ultraviolet spectrum, which is also invisible to our eyes. A small fraction—less than 1% of the total—arrives in the form of even more energetic, potentially damaging, and also invisible x-rays and gamma rays, and as harmless emissions in a wide spectrum of radio waves at the other end of the electromagnetic spectrum.

It comes as no surprise, given the source of light in which life evolved, that the spectral response of our own eyes is so similar to the spectral distribution of radiation from the Sun. This is not the case for stars that are significantly hotter or cooler than the Sun, whose peak emissions are shifted, respectively, toward the violet or the red. Thus we bear in our genes an identifier of the type of star with which we live.

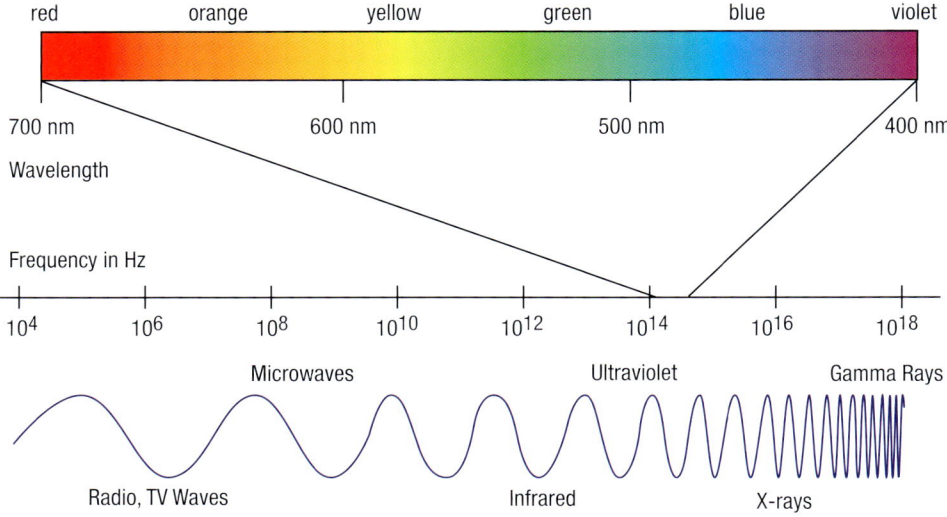

The portion of radiation from the Sun or other source that can be seen by the human eye as visible light (upper bar) in the context of the full range of electromagnetic radiation. Electromagnetic radiation is specified in terms of frequency of oscillation in cycles per second, or hertz (Hz) or alternately, the corresponding distance between crests of adjoining waves (wavelength). Radio radiation, at the far left, includes the lowest frequencies and the longest wavelengths, which can be as long as thousands of miles. At the other end of the spectrum is gamma radiation, with wavelengths shorter than the dimensions of a single atom.

How Constant Is and Was the Sun?

We should expect the Sun's total output of radiant energy to be nearly constant, by virtue of the enormous mass and thermal inertia of the star. Fluctuations in energy production in the core, or elsewhere beneath the surface, should be leveled out in the lengthy and tortuous processes through which heat is passed, hand over hand, through the interior of the Sun.

More gradual changes over billions of years in the Sun's luminosity are not ruled out by this argument. And indeed, what we know of how stars like ours evolve with time makes it almost certain that two or three billion years ago—when the Earth was new and the Sun was young—the Sun was as much as a third dimmer than today. Thus, in the course of its long life, it has indeed changed, and by quite a bit.

We also know from geology and paleontology that much of the surface of the Earth has been covered with water, and with life, for billions of years without interruption. From that crude datum we can say that the Sun has never burned so bright as to cause the oceans to boil during this long period of time. Nor so feebly that all the oceans froze. But neither of these tells us very much regarding the actual range, through time, of solar inconstancy.

What most of us should like to know is how constant and reliable is the Sun today, and whether solar radiation has varied significantly in the past hundred or perhaps a thousand years, and how much it will likely change in the future.

Metered Sunshine

We can now record, from minute to minute, all the changes that occur in the total amount of sunshine that reaches the Earth. But only since 1978, when the first precision instruments capable of making these difficult measurements of total solar irradiance were put in orbit about the planet. Generations of astronomers before that had attempted to answer this oldest of questions about the Sun from the tops of mountains and other high elevation sites, but their efforts were always limited by uncertainties in the variable absorption of solar radiation in the Earth's atmosphere.

We can now monitor as well the course of change in different spectral components, such as the solar ultraviolet or solar infrared radiation. As suspected, all forms of radiant energy that the Sun emits fluctuate on all time scales. The greatest variability is found in the shortest wavelengths—in our receipt of solar x-rays and ultraviolet radiation—and in the longest solar radio waves. The least variability is found in the visible and near-infrared regions of solar radiation. These latter two are also the greatest contributors, by far, to the total amount the Sun emits. As a result, we should expect the total solar irradiance, as measured at the Earth, to vary only slightly.

Continuous measurements taken since 1978 show that in this period the total radiant energy received from the Sun at the top of our atmosphere has indeed changed very little. But "constant" it is not, for it varies from minute to minute, day to day, and year to year, largely in response to the changes we see on the visible, white-light surface of the star. Darker (and hence cooler) sunspots, competing with brighter and hotter areas found around them and around the perimeters of convection cells continually tweak the total energy emitted from the visible hemisphere of the Sun.

Due to these changes on the surface of the Sun, the heat and light delivered at the top of our atmosphere can vary from day to day through a range of about ± 0.3%, and in annual average, from year to year, by about 0.07%. Neither of these solar fluctuations is as large as the everyday fluctuations in the voltage that supplies the lighting in our homes and places of work.

The change of about 0.07%, peak-to-peak, in annual-averaged measurements of solar radiation is among the simplest and most predictable of solar variations, for it marches to the drum beat of the well-known 11-year sunspot cycle. The total radiation released by the Sun into space is—somewhat surprisingly—greatest in those years when there are more sunspots on its surface, at which time increased magnetic activity brightens other regions on its surface. It then decreases by about 0.07% in the six or seven years it takes for solar activity to decline to a minimum level, when fewer sunspots are found.

Though far smaller in magnitude than the effects of daily changes in cloud cover or the annual variation of the Earth's distance from the Sun, this periodic change of 0.07% in total solar irradiance is enough to alter the temperature of both the air and the surface oceans. And indeed, measurable changes of the predicted amount and expected phase (warmer in years of maximum solar activity) have now been found. The impacts of charged atomic particles from the Sun—which also follow the 11-year solar cycle—could leave similar solar marks on the global climate record. How much the outputs of the Sun vary over longer periods of time is today a pressing and as yet unanswered question.

The First Who Saw the Face of the Sun

In Tuscany, in the summer of 1611, Galileo Galilei, then forty-seven, turned his small telescope on the Sun and projected its image on a white screen an appropriate distance beyond the eyepiece, where it was safe to view. At about the same time, three other men, quite on their own and far away, had begun as well to examine the bright disk of the Sun in the same way, like Galileo, with the aid of the newly-invented telescope: Thomas Harriot in England, Johann Goldsmid in Holland, and Christopher Scheiner in Rome. Though they would never meet, these four early explorers of the sky—three scientists and a Jesuit priest—were the first people to look so closely into the face of the star that lights the world.

None of them, including Galileo, claimed to have "invented" the telescope, which had been stumbled upon, purportedly by accident, in a spectacle-maker's shop in Holland a few years before. But it was Galileo—who knew the most about astronomy and optics and the art of reaching and convincing others—who applied the newfound tool most critically and effectively.

His early telescopes, like those of his three competitors, magnified what one could see with the eye alone by about a factor of thirty, and were made of rolled metal tubes not more than an inch or two in diameter and about a yard long. With these truly revolutionary devices Galileo had in 1609 and 1610, from his home in Florence, first looked at the Moon (to find valleys and mountains); at Jupiter (to discover four moons that circled that large and far away planet); at the Milky Way (to find it filled with stars so numerous as to be almost beyond belief); and at Mercury, Venus, Mars, and distant Saturn: which, with Jupiter, were at that time all the known planets in the sky.

What Galileo and the others found on the face of the Sun was a scattering of small dark spots of varied sizes, which were not fixed in place, but seemed to move from day to day across the solar disk. That there were imperfections of any kind on what was deemed to be the Sun's pure white surface may have come as a surprise to all of these first telescopic observers, for sunspots were not known in Europe or in most of the rest of the world at the time. In truth, Galileo and his three contemporaries were not the first to find them. Dark spots on the Sun, seen with the unaided eye, had been reported a long time before, in ancient Greece, and on at least one occasion in medieval Russia. And in China they had been more or less continuously observed, described and recorded since well before the time of Christ.

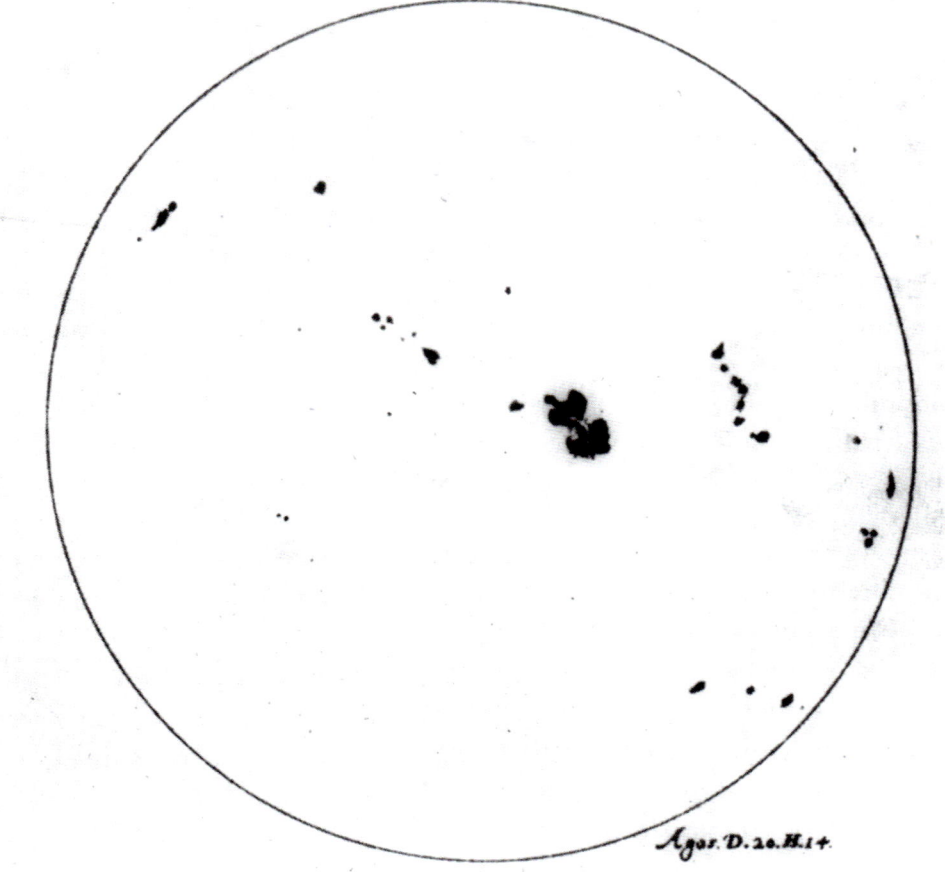

A drawing of the full disk of the spotted Sun made in Florence by Galileo using a small telescope at about 2 o'clock in the afternoon of August 20, 1610, looking much as it does today.

Sunspots that are sufficiently large can be seen without the help of a telescope, at those times or places when the blinding disk of the Sun is sufficiently dimmed. This is possible, for instance, when the Sun is seen through the smoke of a forest fire. One can also do it with the aid of a smoked glass filter that is sufficiently dark; by reflection in water that is smooth and very still; through thick clouds or haze; or at sunrise or sunset when the disk of the Sun is reddened by our atmosphere.

Since well before the time of Christ, sunspots were watched for in one or more of these ways by court astrologers in China (and later in Korea and Japan) who read them in terms of portents, often dire. They also left behind a well-preserved record of their relative sizes and when they were seen, in authorized dynastic histories. But without the help of a telescope, these early Eastern watchers of the sky could not clearly resolve the dark spots which they saw come and go on the face of the Sun, beyond simple, intriguing descriptions such as "as large as a plum", or sometimes "a duck's egg".

Galileo was in a position to make this distinction, and soon did. He concluded that the dark spots were too irregularly shaped to be planets encircling the Sun—as church apologists, including Father Scheiner in Rome, had hastened to propose. They were instead dark features on the Sun's own surface: which was not at all a divinely perfect fire. He also deduced from the manner in which sunspots appeared to move across its face that the Sun rotated, in a period of about 26 to 27 days, and that the Sun's axis of rotation was tilted by a few degrees.

Father Christopher Scheiner and another cleric tracing the image of the spotted Sun early in the 17th century, by projecting the image of a long focus telescope on a sheet of paper. The etching is from Scheiner's massive book on sunspots, the *Rosa Ursina*, published in Latin in 1630.

These robust findings, he concluded, strongly supported the proposition put forward more than half a century earlier by the astronomer Nicholas Copernicus in Poland. Namely, that the Earth was *not* the center of the universe, fixed in space with the planets and the Sun and all the other stars rotating once each day around it, as had long been taught. Instead, it is the Earth that moves around the Sun, spinning on its own axis once each day, while tracing out a year-long course around a tilted, turning and imperfect Sun.

To church authorities—the Christian fundamentalists of his day—these then-controversial views were deemed heretical in that they appeared to contradict the literal interpretation of certain words in the Holy Bible. For this perceived transgression, Galileo—a devout Catholic—was brought to trial in Rome. There he was found guilty of heresy; made to publicly recant what he had said and written; forbidden to teach or publish; and held under house arrest for the remaining years of his life, until the day of his death in 1642.

The Long Watch

Because the photosphere is easily viewed and ever changing, this best known layer of the gaseous Sun has been kept under nearly continuous telescopic surveillance since Galileo's time: watched and monitored through the years by an unending succession of both amateur and professional astronomers. As a result, there exists today in the form of collected descriptions, drawings, and ultimately, photographs, a diary of our star that tells first-hand of changes on the face of the Sun through a span of nearly 400 years.

The white light Sun was regularly watched and carefully documented throughout the 17th and 18th centuries, fueled by the European re-discovery of sunspots in 1611; by Galileo's short book, published in 1613, announcing his findings; by Father Scheiner's massive shelf-bending tome, published in Latin in 1630, which with elegant illustrations, detailed his own daily observations of sunspots; and by the interest of nearly every astronomer of note in the course of those two centuries.

Since early in the 19th century there are observations on record of how the face of the Sun appeared on every day of every year. And in 1848 the long-kept watch was organized into an international effort. Not long after, daily photographs of the solar disk were added to the global patrol; although daily drawings of the spotted Sun—made much as Galileo and Scheiner had done—were still employed for certain purposes for another hundred years and more.

The reasons for keeping so continuous a watch on the Sun changed through the centuries. Initially it was intellectual curiosity and a search for answers: to identify the characteristics and behavior of the dark spots (in Latin, maculae) and the bright patches (called, in part to rhyme with this, faculae, or little torches) that were often seen in close proximity to them when they were observed near the edge of the solar disk. What causes these bright and dark marks on the face of the Sun? What do they tell of the health of the star, or of its inconstancy? In what ways might they affect the Earth?

With the announcement, in 1843, of a more or less regular variation in the number of spots seen on the Sun—rising and falling again in a period of about ten years (later eleven)—the interest in sunspots increased.

Not long after that time, observations of the photosphere made with the new-found spectroscope made it possible to determine the temperatures and pressures and eventually the strengths of the magnetic fields in sunspots and other features of the photosphere. It was these that at last unlocked the secrets of the changing face of the Sun.

Sunspots were *darker* because they were almost 30% cooler than the 10,000° F temperature of the surrounding photosphere. They were cooler because they defined those places on the white-hot surface of the star where strong magnetic fields, emerging from below the visible surface, inhibited the upward flow of heat from the interior of the Sun. Faculae were *brighter* because they were denser and hotter than the surrounding photosphere, and they too were related to the surface magnetic field.

In short, the Sun—though indeed dependable and in the long term, quite constant—was found to be a moody magnetic star on which internally-generated magnetic fields affect the inexorable upward flow of heat from deep within it. Moreover the appearance and demeanor of its radiating surface are entirely driven by the contortions and interactions of immensely powerful magnetic fields: what Oliver Wendell Holmes called *"the maelstroms of the photosphere"* in a prescient poem composed in 1882.

The Sun That We Can See

The *photo-* (or light) sphere is the name given the white-hot glowing layer of the gaseous Sun where almost all visible light and infrared heat originate. As such it is the deepest layer within the star that we can see directly with our eyes, and the deepest from which we can feel solar heat.

Nearly all the red, orange, yellow, green, blue and violet rays that we receive from the Sun—which when seen together, appear to us as white—come from the photosphere. As does the rainbow, which is but white light from the photosphere pulled apart in its passage through mist or rain and then fanned out to form a looming arc of the Sun's original palette of colors.

The same is true for the blue of the sky. It too is white sunlight from the Sun's bright photosphere, scattered over the dome of the day-lit sky by intervening molecules of air. This physical process preferentially scatters the blue and violet rays to lend color to the sky, while letting the rest of the sunlight slip through unseen. But like beauty itself, the blue we see and artists paint lies mostly in the eye of the beholder, for in truth, the daytime sky is predominantly violet. It is only our own retinas—which are less sensitive to violet light—that produce the beautiful blue.

The Photosphere

Today the photosphere is watched and monitored around the clock from solar telescopes in solar observatories around the world—and from the vantage point of space—with ever-better clarity and magnification.

One of the purposes for this close and continuing watch is to improve what we know of how photospheric features, including sunspots and faculae, alter the amount of heat and light that we receive each day from the Sun. Another is to refine our knowledge of the causes and sources of explosive solar flares, eruptive prominences, and coronal mass ejections, all of which originate in photospheric magnetic fields.

Behind each of these activities is our growing need to anticipate those solar changes that impact an ever more crowded and more technical world. Toward these ends, each improvement in our ability to look more clearly into the face of the Sun is a step in the right direction.

From his back yard in Florence and using his best telescope, employing the glass lenses he had ground himself, Galileo in 1611 was able to distinguish individual features on the Sun that were as small as about 2000 miles across—roughly the size of the Moon, or the continent of Australia—but nothing smaller than that. With early improvements in the telescope, Father Scheiner in the 1620s was able to resolve details within sunspots that were about two times smaller than what Galileo had been able to see, or about the size of Alaska. Fifty years ago, although telescopes were by that time much improved, what could be seen on the Sun from even the best observing sites was limited by turbulence and irregularities in our own atmosphere to features that were no smaller than the state of Texas, roughly 500 miles across.

Revolutionary advances in optics and technology have followed since then. Solar telescopes in space operate far above the blurring atmosphere and offer

The solar disk seen in white light from the SOHO spacecraft on October 28, 2003, showing (near the bottom in this figure) an unusually large and complex sunspot group which give birth to fast-moving coronal mass ejections and some of the largest solar x-ray flares ever recorded. The area encompassed by the large group of spots equaled that of at least 15 Earths, surpassing the size of Jupiter.

as well the ability to see the Sun in ultraviolet and x-ray wavelengths, not visible from the ground, as well as the ability to continuously observe the Sun's chromosphere and corona.

Ground-based solar telescopes, utilizing computer-controlled adaptive optics that automatically and continually compensate for irregularities in the air above them, are now able to distinguish features as small as 50 miles wide on the disk of the Sun. With the aid of this powerful new technology, the *Advanced Technology Solar Telescope (ATST)*—now under design at the National Solar Observatory—will allow tomorrow's solar observers to see and examine features as small as 12 miles in scale on the face of Sun, the size of downtown Denver, from 93 million miles away.

Sunspots

Far and away the best known features of the Sun are the black spots—ranging in size from too small to see with any telescope, to twice the size of the Earth—which are seen on its face in varying numbers on almost every day of every year. At no time are sunspots scattered over the whole Sun. They appear instead in restricted belts of solar latitude that gradually migrate from high solar latitudes toward the Sun's equator in the course of every 11-year sunspot cycle.

They also are most often clustered in separated groups, like flocks of birds. Individual spots are born small, grow in the course of their lifetime of days to weeks and sometimes months, and then fade from view. As Galileo noted, they all move systematically from the *left* or *eastern* edge of the Sun toward the *right* or *western* limb, carried along like autumn leaves on a brook, as the star turns slowly on its axis, completing each solar rotation in about 27 of our days.

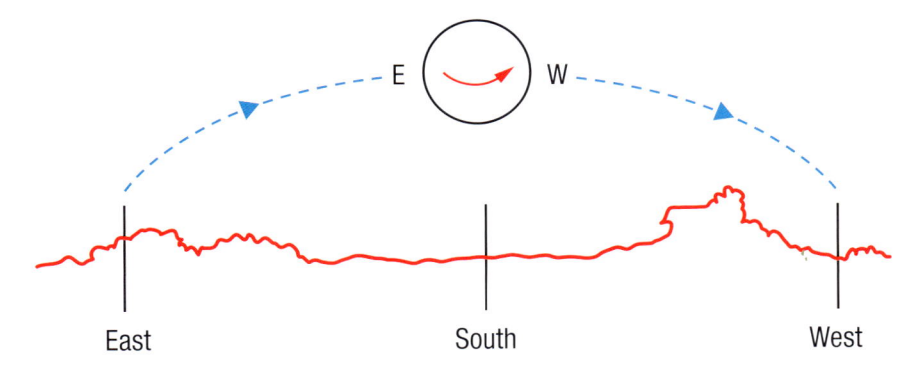

A schematic view of the horizon showing the daily east-to-west path of the Sun through the sky (as a dashed blue line), the direction of the Sun's rotation about its axis (as a red arrow) and the east and west limbs of the solar disk.

When seen through a small telescope of the sort that Galileo employed, most larger sunspots are found to consist of a darker and hence somewhat cooler central region—the umbra, meaning shadow, or shade—surrounded by a less dark fringe—the (partially shaded) penumbra. A closer examination of the penumbra reveals that it is made up of what look a lot like buttonhole stitches sewn around the perimeter of an umbral hole. In reality, these penumbral features are towering arches of heated solar gas that flow both into and out of the umbra of the sunspot, constrained by curved lines of magnetic force.

Any and all unsolved mysteries regarding sunspots—why they appear darker and take the varied shapes they do, where they come from, and what purpose they serve—were swept away forever when it was demonstrated, early in the last century, that sunspots are the sites of highly concentrated magnetic fields.

To summarize: sunspots are darker than the surrounding disk of the Sun because—though very hot by terrestrial standards—they are considerably cooler than the rest of the photosphere: about 7000° F instead of 10,000. They are cooler because they mark the places where lines of magnetic force, emerging from the churning interior of the star, are concentrated and clustered together, forming a kind of magnetic plug which at those places inhibits the upward flow of heat.

The penumbral arches delineate the boundary where the tightly packed vertical field lines in the umbra of a sunspot begin to spread and fan out, like the bloom on a lily.

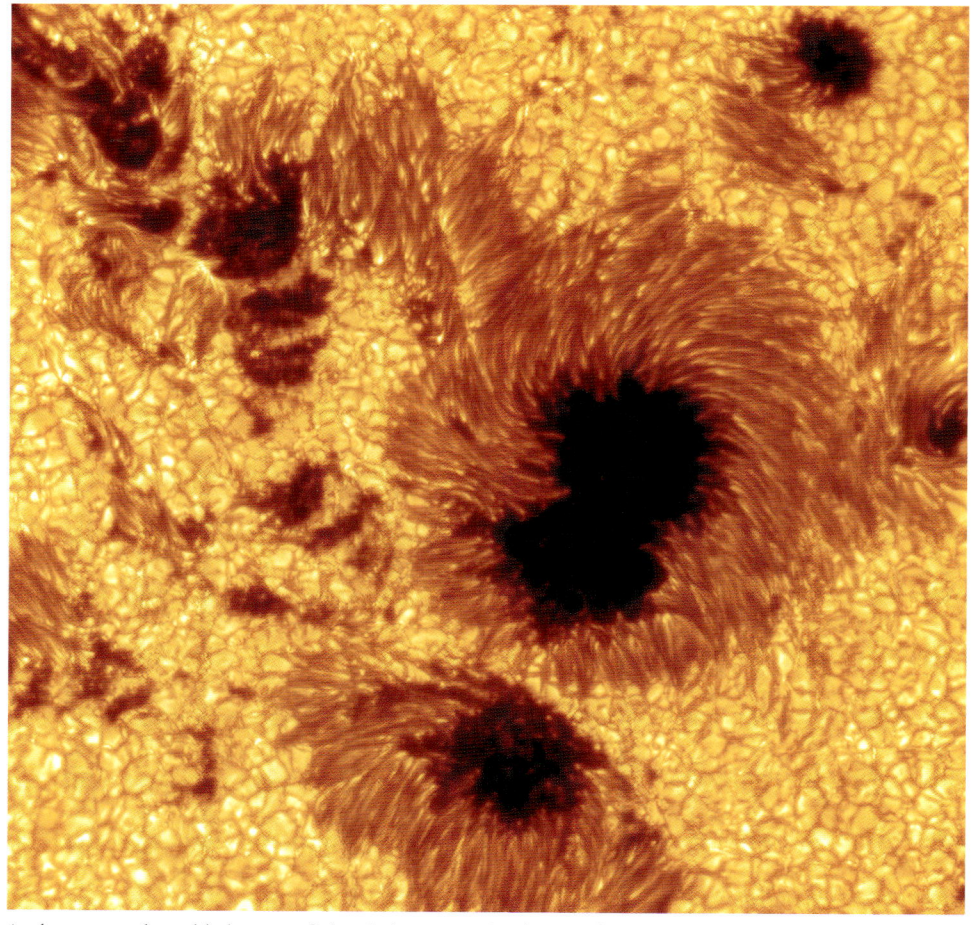

A close-up, color-added view of the dark centers (umbrae) of sunspots; the less-dark, magnetically-formed penumbrae that generally encircle them; and the ubiquitous pattern of close-packed tops of the convective cells that characterize the entire photospheric surface of the Sun. This unusually clear image was made from the mountain-top La Palma observatory in the Canary Islands using the Swedish 1-meter high-resolution solar telescope, which employs automatically-adjusted optics to continually compensate for the blurring effects of turbulence in the Earth's atmosphere.

Bright Faculae

Brighter (and hence hotter) regions on the photosphere, called faculae, are another consequence of solar magnetic fields. A dermatologist would probably describe them as a rash of luminous blotches on the face of the Sun: larger and far more irregular in shape than sunspots; not as much brighter than the surrounding photosphere as sunspots are darker; but more extensive and spread over a larger area of the solar disk. They are also more ephemeral—living but a day or so—and ever changing. The largest are found in the close vicinity of sunspots, and like them the area of the Sun covered by faculae rises and falls in step with the 11-year sunspot cycle.

Due to their lower contrast, faculae are not easily seen with a simple telescope. This did not stop Galileo and Scheiner and the other early 17th century astronomers from finding them when these subtle brightenings were near the edge (or limb) of the solar disk, where they are far more apparent.

The faculae we see near the *left* or *eastern* edge of the Sun are carried by solar rotation toward the center of the solar disk where their apparent brightness—relative to the surrounding photosphere—rapidly fades, and they soon drop out of sight. A week or so later when the same facular areas approach the *right-hand*, *western* limb of the Sun they once again become more apparent. The reason why faculae can be seen at the edge but not the middle of the solar disk follows from their towering form and the background against which they are seen.

Faculae consist of vertical columns of magnetically-constrained gas that is hotter and hence brighter than the photospheric surroundings through which it flows. Because of this, faculae on the face of the Sun increase the total amount of heat and light that the Sun emits and we receive. The sunspots that labor beside them to hold back the outward flow of energy from within the star work in the opposite way, to diminish the total, in an ongoing Yin and Yang relationship that is probably as old as the Sun itself.

It never ends in a draw. In terms of day-to-day changes it is sunspots that hold the upper hand—and a heavy one—turning down the Sun's output of energy on any day by as much as several tenths of a percent. But in the longer run—when averaged over months to years—the faculae prevail, aided in part by other bright features of smaller scale more uniformly distributed on the solar surface that are even more difficult to see.

Both the number of spots and the prevalence of faculae rise and fall together in the course of the 11-year solar activity cycle. Since it is the bright features (and not the sunspots) that play the heavier hand, the Sun *brightens* and the heat and light we receive from the Sun systematically *increases* in the maximum years of the cycle. When there are fewer and fewer sunspots—in years of lower solar activity—the total radiation *decreases*. And although the change is less than 0.1 percent, it is enough to perturb our climate system, in part through persistence, by pushing in the same direction for months or years at a time.

Beneath the Shining Surface: The Bubble Machine

The photospheric background against which sunspots appear is itself highly structured: made up of a closely-packed honeycomb of brighter elements, each bounded by darker lanes. This intriguing, almost geometrical pattern, first described more than 200 years ago, covers the entire surface of the Sun, from pole to pole. Because the close-packed elements looked something like small pellets or grains, they were initially called granules, and the overall pattern the solar granulation.

Today we know that what we see as the granular, mottled photosphere is in fact the top surface of a deep convective layer that fills the entire outer third of the Sun's interior. Within this vast and ever churning region, some 150,000 miles deep, intensely hot gases heated by the Sun's nuclear furnace, far below, are made more buoyant and hence propelled upward—like bubbles in a heated kettle.

After continual jostling, interaction and exchange, some make it to the surface of the star. There in the crowded company of billions of other glowing granules they cool by giving up some of their energy in the form of radiant heat and light. Once cooled and therefore heavier, the hot gases within them sink down once more beneath the surface, to be reheated and fight their way to the top again.

The heated cells take their honeycomb form in a variety of shapes and sizes. The smallest we can yet discern are about 100 miles across; the average, twice the size of Texas; and the largest are about as big as the continental USA. But life at the top is a fleeting thing for any of them, large or small. Once there, they have but a few minutes in the limelight before up-and-coming others crowd into their place.

To help discharge their cargo of heat while at the surface of the Sun, each convective cell is internally stirred by patterns of flow that circulate hot gases upward in the center, outward toward the perimeter and then back down again at the sides. And while examples of churned mixing of this kind come readily to mind—as in a kitchen blender, or when stirring paint in a can—none of them works as fast or frenetically as these gigantic mixers on the Sun. Within each solar convective cell, the hot streaming gases are carried up, across, and down not like batter in a bowl but at the dizzying speed of a mile each second: faster than a speeding bullet.

Lifting the Veil: The Unseen Sun

Almost all that is known of the Earth's interior has been obtained indirectly, without ever seeing or sampling it. Particularly valuable is the application of seismology: the science by which one can probe the actual interior of the Earth by the way sound waves—initiated at the surface—are bent and reflected back again.

The study of the hidden interior of the Sun—of what lies beneath the photosphere—has made use of the same techniques, which in this application is called helioseismology: the study of the solar interior based on observable oscillations on its surface.

In terrestrial seismology, the disturbances that initiate sound waves include earthquakes and other natural tremors of opportunity as well as man-made disturbances that are set off for this purpose. In the case of the Sun the force that induces sonic waves is the ever-present piston-like up and down movement of material in the solar convection zone: like rambunctious children jumping up and down with all their might on the bed in a motel room.

In response to this incessant hammering, the entire photosphere is made to oscillate, in distinctive patterns of undulating waves that slosh slowly up and down, in a regular period of about five minutes. Some of this created energy is thought to heat the chromosphere and corona. Another part is directed back into the interior of the star, in the form of sound waves. These waves, as they pass through the nether world of the Sun, are refracted (or bent) by differences in the internal properties of the solar interior.

The bent paths of these solar sonic waves—like those that are employed to probe the inner Earth—ultimately take them back to the surface again, where they perturb the natural oscillation of the photosphere. It is these subtle differences in solar oscillation, measured and compared from point to point on the surface of the Sun, that are now used to reveal many of the secrets of the innermost Sun. For these purposes data are taken both from spacecraft and from a dedicated around-the-world network of automated ground-based stations.

The application of these techniques have confirmed the existence of a deep convection zone and plumbed its very bottom, 150,000 miles deep. They have helped illuminate what happens just beneath a sunspot, and the depth at which the bundled magnetic field lines that give it form and function begin to block the upward flow of heat within the Sun. And they have shed a bright new light on the origins of the Sun's varying internal magnetic field and the mechanisms within the Sun that control the birth, the places of birth and the regularly-varying numbers of sunspots and related features of solar magnetic activity that affect our own lives in down-to-Earth ways.

☉ ☉ ☉

The most fundamental finding from helioseismology, thus far, is probably the clear-cut information it provides on how the inner Sun rotates, and how the rate of rotation changes with depth and latitude.

There is no reason to expect that the squishy, gaseous Sun should rotate like a spinning bowling ball—at the same rate everywhere, inside and out. Indeed, it has been known since Galileo's time that different latitudes on the surface of the Sun rotate at different speeds: completing a turn about five days faster at the equator than near the poles. Were the Earth to follow that solar recipe, the number of hours in a day (say, from noon to noon) would depend upon one's latitude: with each day in Anchorage or Saskatoon several hours longer than one spent on the beach at Waikiki.

Helioseismology has shown that the rate of rotation of the Sun also varies considerably with depth.

The Sun's convection zone—much like the visible surface of the star—rotates much faster at lower solar latitudes than it does nearer the poles. At the equator, it takes about 27 days for matter on the solar surface and in the convection zone beneath it to complete one full turn around the rotational axis of the Sun. Near the poles it takes almost half again as long to make the trip.

One consequence of this *internal* differential rotation is the distortion and twisting of sub-surface magnetic field lines. Magnetic lines of force within the Sun—which would otherwise run from pole to pole—are by differential rotation twisted into toroidal magnetic fields which lie in planes *perpendicular* the Sun's axis of rotation. This is one of the forces that bring internal magnetic fields to the surface to appear as sunspots and other manifestations of solar magnetic activity.

A second level at which differential rotation occurs is deep within the solar interior, at the depth where the bottom of the convection zone comes in contact with the hotter, deeper, denser, and dynamically different radiative zone that lies beneath it. At this level and below, solar material rotates as though it were indeed a solid inner sphere, in twenty-seven days—which is at high latitudes much faster than the convection zone that lies immediately above it. Nearer the equator it spins more *slowly* than the overlying convection zone.

The spherical Sun sectioned to show what lies beneath the chromospheric surface of the star. Within the bright central core (white) atomic hydrogen under extreme pressure and temperature fuses to create atoms of helium with an accompanying release of energy, which will ultimately escape from the photosphere. Surrounding the core is a vastly larger radiative zone (light orange) through which solar energy is passed outward from atom to atom by the process of radiation, like warmth from a room heater. When the energy passed in this way gets about two-thirds of the way to the surface of the Sun, it is transmitted further upward by the mechanical motion of circulating cells within the tumultuous convective zone (dark orange).

The result is continuous slippage and shear within a thin and troubled layer, called the tachocline, which separates these two interior shells.

It is also thought that the tachocline may be the layer where all solar magnetic fields are conceived, and as such, a fundamental piece of the basic mechanism—known as the solar dynamo—which drives eleven year and longer fluctuations in solar magnetic activity.

The Sun's Chromosphere and Corona

Although it looks that way when seen with the unaided eye or viewed through a simple telescope, the well-defined edge of the Sun is not its outer boundary. The familiar white photosphere is only the brightly shining core of a far larger star. Were the rest of the Sun (the chromosphere and corona) as bright it would appear more than ten times larger, and bounded by an ever-changing ragged and asymmetric shape.

The chromosphere is the relatively-thin, tenuous layer of the Sun that lies just above the photosphere. The name *chromo-* (or color) sphere comes from how this layer in the solar atmosphere appears during a total solar eclipse, when for but a few fleeting moments—just before and just after the moving Moon completely covers the full disk of the Sun—we see it edge on, in the intense red light of hydrogen.

Just above it is the more tenuous and far more extensive corona: literally the crown of the Sun, which is so completely different from the rest of the star that it seems like a ghostly other-world appendage. Contributing to this impression is the fact that this very real part of the Sun is so rarely seen. For most people, the only chance will come if they seek out or happen to be caught within the small, speeding shadow of the Moon (about 100 miles in diameter) during a total eclipse of the Sun, which last at most seven minutes. Because this happens at a given place, on average, only about once in 400 years, almost all who through time have lived on this planet never saw it, as most people today probably never will.

The chromosphere and corona, in part because they are more tenuous and diffuse, are so much less bright than the disk of the Sun that under normal conditions they are blocked from our view: much as the bright headlights of a close-approaching truck bar us from seeing the vehicle itself. And though a part of the Sun, they obey a quite different set of rules. Both the chromosphere and corona are non-uniform and highly structured, ever changing, often explosive, and shaped by magnetic forces into forms of awesome beauty.

☉　☉　☉

What is most surprising is that these outer reaches of the star are far hotter than the glowing photosphere that lies just below them, which flies in the face of all common experience and intuition. We expect the temperature in the vicinity of an internally-heated object, like the Sun or a cast-iron stove, to drop, not rise, as we move farther away from the source of its heat. But because of their make-up, these outer layers of the Sun are not bound by these laws of thermodynamics. Other, non-radiative sources of energy must be involved, although identifying them with certainty remains a challenge.

What heats the chromosphere and corona to such high temperatures? The auxiliary source of heat is most likely found in either the intense magnetic fields that thread these layers of the solar atmosphere, and/or the mechanical energy deposited at the base of the chromosphere from the relentless pounding of convection cells: the same source that imposes the undulating, five-minute patterns of oscillations in the photosphere. What is not fully understood is how mechanical energy deposited at the base of these cloud-like extensions of the Sun can heat the chromosphere and corona so far above it so efficiently.

From the core of the Sun outward the temperature falls steadily, mile after mile, through a distance of almost half a million miles: from about 29 million degrees Fahrenheit in the nuclear furnace to about 10,000°F at the radiating surface of the photosphere. For another few hundred miles above this visible boundary the temperature continues to coast downward until, about a quarter of the way through the thin overlying chromosphere, it has fallen to about 7000°.

There the temperature of the Sun abruptly reverses its long and leisurely fall. Within a very short distance above that point the temperature of the thin chromosphere jumps to about 20,000° F. Just above the chromosphere, in an even thinner transition zone that separates it from the corona, the temperature is ten times higher. And not far above that, in the new and different world of the solar corona, temperatures are measured in millions of degrees.

☉ ☉ ☉

The photosphere consists of both neutral and ionized atoms, as well as some molecules. With each increase in temperature—as one moves through the chromosphere and transition zone and into the corona—more of the atoms and molecules of any and all chemical elements are stripped of more and more of their electrons, producing more ions, carrying a positive charge, and free electrons with a negative one.

A collection of charged particles of this kind—common in the atmospheres of stars—which contains roughly equal numbers of electrons and positively charged ions, defines a fourth state of matter: not the solids, liquids or gases of our ordinary experience, but highly ionized atomic particles—or plasma. In this altered state it has sold its soul to magnetism, and is now subject to every whim of the lines of strong magnetic forces, rooted in the convection zone and photosphere, that twist and weave their way through the atmosphere of the

Sun. In truth, above the photosphere all that is left of the star is controlled, in form and in function, by the Sun's magnetic field.

These magnetic lines of force confine and shape the solar plasma, molding it into the many forms that distinguish and decorate the outer atmosphere of the Sun. In the chromosphere, some magnetic field lines corral and organize the hot solar plasma into a pattern of close-packed super-sized granulation cells, 10,000 to 20,000 miles across and but 1000 miles deep, covering the entire Sun. The super-granulation cells that make up this overlying chromospheric network are bounded at their edges by hedgerows of tall, magnetically-formed spicules—or spikes—that carry hot confined plasma upward into the corona.

The largest and longest lived of the many magnetically-sculpted features in the chromosphere are the protuberances, now called solar prominences, that protrude in a variety of shapes high into the far hotter corona. They are for the most part formed of magnetic loops, the most spectacular of which appear above the edge of the Sun as towering arches—some active, some inactive or quiescent—that extend from about 30,000 to as much as 250,000 miles above the photosphere. In form and grandeur these large loops of cooler plasma look a lot like the Gateway Arch—were it painted a fiery red—that soars above St. Louis on the banks of the Mississippi: or croquet hoops of colossal scale, some tall enough for mighty Jupiter, the largest of the planets, to roll quite easily through them.

☉ ☉ ☉

Much of the lower corona is made up of magnetically-formed arches whose foot-points are rooted in the photosphere in regions of opposite magnetic polarity, most often in sunspots. In solar images made in the x-ray region of the spectrum, where radiation emitted by a million-degree plasma is best seen, the lower corona of the Sun is so heavily stitched with these magnetic loops that it looks a lot like one side of a Velcro fastener, awaiting the closure that will never come.

Coronal magnetic fields at low and middle latitudes shape the outward-flowing coronal plasma into tapered forms, called coronal streamers, which extend far into interplanetary space. Against the darkened sky of a total eclipse of the Sun, these graceful extensions of the outer solar atmosphere are made visible to us, looking very much like the petals of a white dahlia.

We see these and other features of the white and ghostly corona at times of a total solar eclipse not by their own weak emission of light but by the scattering or redirection of white light coming upward from the photosphere.

The white clouds we see in the sky are visible to us through a similar process: in this case the scattering of white sunlight by the microscopic water droplets which we would otherwise not see. The same scattering process illuminates nighttime fog in the bright headlights of an automobile, and droplets of mist that appear as a spherical glow around a street light.
In the corona the scattering particles are electrons which are particularly efficient scatterers. What we do not see in the corona are the protons and ions which are also present. Since these atomic particles are far heavier than electrons, they respond less readily to incident light from the photosphere. Like the Moon, or the water droplets that make up clouds, we see the scattering electrons only in the reflected light of the photosphere. Like clouds on Earth the outer corona is white because the photosphere is white. Were it purple or green the corona would be also.

Thus, during the few rare minutes of a total solar eclipse we are allowed to see—from 93 million miles away—the two ingredients of which the corona is made: electrons, which are so small that it would take 10^{30} of them to weigh 1/3 of an ounce; and the lines of force of the Sun's magnetic field, which like steel girders give shape and form to all coronal features.

Where coronal streamers appear and how fully they surround the central disk of the Sun is determined by the location and strengths of magnetic fields on the surface of the star. Because of this, the appearance of the corona, however it is observed, changes considerably from day to day and systematically from year to year with changing levels of solar activity.

The high latitude corona has its own distinctive appearance. There, the radial extension of the Sun's polar magnetic field arranges the coronal plasma into a crown of spreading polar plumes that encircle the poles, as though to guard them, like illumined palisades.

TEMPERATURES BENEATH, AT, AND ABOVE THE VISIBLE SURFACE OF THE SUN

LOCATION	TEMPERATURE IN DEGREES (F)
Nuclear fusion interior	29 million°
Photosphere	11,000°
Sunspot Umbra	7000°
Low Chromosphere	18,000°
Transition zone	180,000°
Inner Corona	2 million°
FOR COMPARISON: Industrial Blast Furnace	2000°
Oxy-acetylene Flame	6300°
Iron-welding Arc	11,000°

How We See the Corona and Chromosphere

Two constraints keep us from seeing the Sun's outer atmosphere under ordinary viewing conditions. The first is the great difference between the brightness of the corona and underlying chromosphere and the adjacent photosphere, due to the immense difference in the density of matter in these outer and more ethereal layers. The chromosphere is ten thousand times dimmer than the photosphere when seen in the visible spectrum. The corona, in the innermost and brightest regions, is almost a million times dimmer than the brilliance of the adjacent solar disk.

In addition, since the density within the corona decreases with height, its own brightness rapidly fades with increasing distance from the limb of the Sun, until it approaches the limits of practical detection: soon ten million, then a hundred million, then a thousand million times dimmer than the bright photosphere.

The second obstacle is the competing brightness of the sky itself, which no matter how clear and blue, is far brighter than the solar corona, and particularly in the sky immediately surrounding the Sun, which is precisely where the corona appears. Nor can we hope to catch a fleeting glimpse of the corona or the chromosphere just as the Sun dips below the horizon, or as it slides behind a cloud or the edge of a barn, for the brightness of the daylit sky around it will still defeat our every try.

Until fairly recently the only way to see the Sun's corona was the way it was first discovered: in the fleeting moments and within the restricted geographical bounds of a total solar eclipse. A total eclipse occurs, somewhere on our planet, about every year and a half; but as noted earlier, at a particular place, be it London or Topeka, Kansas, only very rarely.

During a total solar eclipse, the Moon—a quarter of a million miles away—blocks for a few minutes the bright light of the photosophere before it can reach and illuminate the atmosphere of the Earth. Through these rare windows of opportunity—most often in far away places with strange sounding names—we are allowed to see the outer atmosphere of the Sun in all its glory, against the background of a sky turned suddenly dark and filled with other stars.

For a few seconds of these few minutes, we can also catch a fleeting glimpse of the chromosphere as well, as a thin, red-colored layer just beneath the corona, as well as any large prominences that happen to lie at the edge of the Sun. Unlike the corona, the denser and cooler chromosphere and prominences shine by their own weak radiation, which is dominated in the visible spectrum by the same

An exquisite photograph of the solar corona, also taken in India during the total eclipse of the Sun that happened to cross that then independent nation on August 14, 1980. Magnetically-formed, bulbous-based coronal streamers—long described as resembling the petals of a dahlia—reach outward into space from both low and high solar latitudes, as is typical at the maximum phase of the 11-year activity cycle, when this image was made. Coronal streamers of the type seen here can be torn free of the Sun in the course of violent eruptions and thrown bodily outward as gigantic blobs of ejected coronal plasma into interplanetary space.

red-colored emission from atomic hydrogen that colors the aurora borealis and australis. The chromosphere, brighter than and not as elusive as the corona, can also be observed on a routine basis, and over the entire disk of the Sun, without the need for an eclipse: using a spectrograph or narrow optical filters designed for this purpose. These optical devices accomplish this feat by isolating the sunlight in narrow spectral lines that are particularly bright in the chromosphere, such as the red emission line of atomic hydrogen mentioned above.

Limited observations of the inner corona, made without the need for an eclipse, have also been possible for more than fifty years, employing specialized telescopes called coronagraphs. These employ optical techniques to create, in effect, an artificial eclipse of the Sun within the telescope itself, and have been operated for decades from mountain-top observatories where the daytime sky is clearer and a darker blue. But until they could be carried into space—where the Sun is always shining and the sky is always black—coronagraphs on the ground could never replicate, in either extent or spatial detail, what one saw of the corona in the brief moments of natural eclipses of the Sun.

Today, coronagraphs carried on spacecraft monitor and observe the Sun's outer, white-light corona around the clock, day after day and year after year. They also provide images that in many ways surpass those obtained at solar eclipses or with coronagraphs on the ground. And since they operate continuously, spaceborne coronagraphs are able to identify and follow changes in the corona that elude the still photographs taken with eclipse cameras. In but one day a coronagraph mounted on a spacecraft above the Earth's atmosphere provides more minutes of coronal observing time than what could have been accrued at all the total eclipses of the Sun in the last 1000 years.

Extant photographs of the solar corona taken at eclipse—which began within a decade after Louis-Jacques-Mandé Daguerre's remarkable invention in 1839—comprise a valued collection of *snapshots*, taken a few years apart, of the evolving form of the white-light corona. But the high-resolution images from space—which became available in the early 1970s—provide a continuous *movie* of the outer corona that catches every change, fast or slow, including particularly the release and expansion of huge chunks of coronal plasma, called coronal mass ejections or CMEs that can directly affect the Earth.

The 40 ft-long telescope of California's Lick Observatory set up near Bombay at Jeur, India to observe the total eclipse of the Sun on January 22, 1898. The cumbersome, long-focus telescope used a 5-inch lens to project an image of the Moon 4 inches in diameter and the corona up to 10 inches across, on 18 x 22 inch glass photographic plates. A series of these were adroitly inserted and removed from the focal plane by hand, one at a time, in the dark of the eclipsed Sun, at this site within a deep pit dug beneath the striped tent. The telescope, called "Jumbo," was taken to fifteen total solar eclipses on six continents from 1893 through 1932, each affording an opportunity of but a few minutes to see the fleeting corona of the Sun.

⊙ ⊙ ⊙

Direct observations of the Sun from space has brought about an equivalent revolution in our understanding of the inner corona and the chromosphere, through round the clock observations of the Sun in the invisible ultraviolet and x-ray light which these regions emit. Since these incoming solar rays are absorbed in the high atmosphere, we are prevented from seeing the x-ray or ultraviolet Sun from the ground.

Telescopes in space that are capable of imaging the Sun in x-ray radiation see the inner corona, more than a million degrees hot, in the light of its own radiation. The familiar x-ray images employed by your doctor or at the airport differ in that they are not pictures of the *source* of x-radiation (in this case a high-voltage vacuum tube deep within the apparatus) but *shadow pictures* of your opaque bones or carry-on luggage, illuminated from behind by a point source of x-ray radiation.

X-ray images of the inner corona offer the great advantage of displaying the entire visible hemisphere of the Sun, as in a bird's eye view from above, as opposed to the more limited edge-on views seen at an eclipse. Moreover, sophisticated x-ray telescopes in space provide these images in extremely fine detail, like high-definition television.

Similar observations of the Sun made from space in the extreme-ultraviolet reveal the upper chromosphere and the transition zone above it in full hemispheric coverage and in fine detail.

As noted earlier, the temperature of the Sun's outer atmosphere rises rapidly with height. With increasing temperature the maximum emission from the region shifts toward shorter and shorter wavelengths. Radiation from the 10,000° photosphere, for example, peaks in the yellow portion of the visible spectrum; from the million degree corona it has shifted all the way into the far ultraviolet and x-ray region.

Thus by a judicious selection of wavelength, including the choice of specific spectral lines, one can isolate and observe different layers of the solar atmosphere, from the photosphere through the chromosphere and transition zone into the low corona: much as adjusting the focus on a pair of binoculars allows one to observe objects that are closer or farther away.

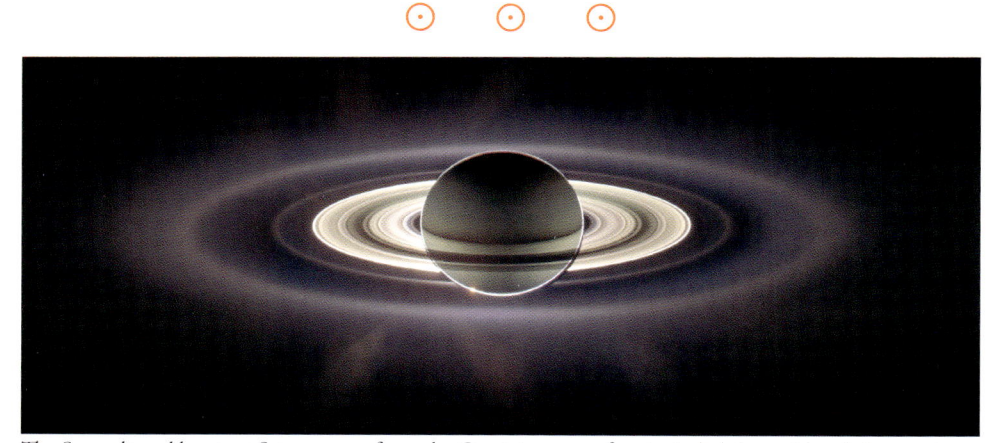

⊙ ⊙ ⊙

The Sun eclipsed by giant Saturn, seen from the Cassini spacecraft as it circled that cold and distant planet in 2006. The bright narrow ring that goes fully around the circular Saturn is photospheric light diffracted from the occulted Sun which lies behind it. Because the Sun is more than nine times farther from Saturn than from the Earth, it would appear much smaller in Saturn's sky than in our own; in this view, made from a relatively short distance from the planet, Saturn's disk appears far larger than the distant Sun hidden behind it. The dimly-lit night side of the planet is here illuminated by indirect sunlight reflected from its many extensive rings, some of which were not found before this image was obtained. Far in the distance, hardly visible and little larger than the specks of other stars is our own small pale blue planet. It can be found in the dark space between Saturn's outermost bright, fuzzy ring and the first, thin dimmer ring, closer to the latter, and at an angle corresponding to about 9:30 on the face of a clock.

A computer-aided depiction of the outward flow of charged atomic particles from the Sun in the solar wind, shown above an actual image of the hot upper chromosphere, made in the far ultraviolet.

THE SOLAR WIND AND SOLAR VARIABILITY

The Solar Wind

A consequence of the extremely high temperatures found in the corona is a corresponding increase in pressure—as happens in a covered pan heated on the stove. In this case there is no lid, and the heated corona expands freely outward, pushing against the yielding, lower pressure of the interstellar gas.

This thermally-driven flow of ionized, charged particles—known as plasma—constitutes an ever present wind that carves out a cavity, called the heliosphere or realm of the Sun, in the surrounding interstellar medium.

Thus, in addition to heat and light the Sun also releases a continuous flow, called the solar wind, of atomic particles—protons, neutrons, electrons and ions of all the solar elements—that expands outward, night and day, in all directions everywhere. This never-ending flow of coronal plasma is of considerable significance for the Earth and the other planets. And so it has been for billions of years.

Notably absent in the solar wind that reaches the Earth, are neutrons: the fundamental atomic particles with neither positive nor negative charge which with protons are the building blocks of atomic nuclei. But although neutrons are present in great abundance on the Sun and are driven off with other particles in the solar wind, the lifetime of any one of them is so ephemeral—lasting, typically, no more than a few minutes before it decays—that most of them are gone by the time they reach the orbit of Venus.

The other particles continue on and on, as the Sun's principal contribution to the universe, filling every crevice of the heliosphere—beyond the dimmest reach of its light beams—with windblown seeds of itself. And they keep on coming, in seemingly endless supply, as from a magical dandelion. In but one second of this continual outpouring, the wildly extravagant Sun gives a million tons of itself away. And in the course of a year, almost 10^{13} tons.

With little to slow it down, the high-speed solar wind reaches the orbit of the Earth—after traveling 93 million miles—in but two days; and the slower wind in about four. At the Earth the average velocity of the streaming solar wind is about 225 miles per second. The fastest streams that reach the Earth blow at about 500 miles per second: a *thousand times* faster than a speeding bullet.

When it arrives at the Earth—however fast it moves—the solar wind is in terms of mass mostly nothingness, containing about 130 protons in a cubic inch, compared to the zillions of atoms and molecules in the air we breathe. And because it is so diffuse, were the fast moving solar wind plasma to blow against our face it would feel far more like a baby's breath than the blustery winds we commonly sense at the surface of the Earth. Even so, the solar wind plasma at the Earth is about 50 times more densely packed than the ambient conditions it will eventually meet outside the heliosphere.

Once past the orbit of the Earth, the solar wind plasma continues streaming outward at constant speed for more than a hundred times our distance from the Sun: well past the orbits of Mars, Jupiter, distant Saturn, and far away Uranus, Neptune and little Pluto.

Somewhere well beyond the planets it will reach the cold and darkened limits of the Sun's domain: the as yet unexplored boundary that marks the end of the heliosphere. Here, a long way from home, its initial force is so depleted by dispersal over so vast a volume of space, that the solar wind loses the upper hand. The weakened pressure of outward streaming plasma no longer exceeds that of similar but alien particles arriving from countless other stars. There, at the heliopause the solar and stellar winds will meet, though because of their differently oriented magnetic fields, they will seldom truly mix.

☉ ☉ ☉

Since it is composed of charged particles, the streaming solar plasma travels outward from the Sun still bearing the embedded (or "frozen in") magnetic signature of the place from which it came, carried with it like a trailing ribbon still tethered at one end to the Sun. Because the surface magnetic field of the Sun is organized into discrete patches of either positive (outward-directed) or negative (inward) polarity, the magnetic earmark of its place of origin is preserved in each stream.

The solar wind that streams outward in this way from equatorial regions of the Sun is divided by these differences in polarity into spatially discrete portions or sectors, which expand in width with distance from the Sun. Although the plasma flows radially outward, the open field lines that it brings with it—still attached to the solar surface—are shaped into a curved spiral by the 27-day rotation of the Sun. As the Sun rotates, sectors of different magnetic polarity in, above or below the equatorial plane, are swept across the Earth, like a spray of water from a revolving garden hose, exposing our own magnetosphere—for a week or more at time—to first one and then the other polarity, in a varying sequence of slow and fast plasma streams.

At those times when the direction of the imbedded solar field opposes that of the Earth, we are more vulnerable to the impacts of incoming solar particles. Under these conditions, where the onrushing plasma makes contact at the Sun-facing "nose" of the magnetosphere the opposing magnetic lines of force merge and connect. There, through a process peculiar to highly-conducting plasmas called magnetic reconnection, magnetic energy from the Sun is efficiently converted to kinetic energy in the Earth's magnetosphere.

When this happens, the drawbridge is down and some of the energy of the onrushing solar plasma makes its way into our own magnetic field, with consequences that can perturb and disrupt conditions on the surface of the Earth.

As noted earlier, it takes several days for the solar wind plasma to reach the Earth. This means that by the time it arrives here, its place of origin on the Sun has already been carried by solar rotation more than halfway toward the right-hand or western edge of the Sun, soon to disappear from our view.

Were we standing on Jupiter—five times our distance from the Sun—the time between release and arrival would be that much greater. There, by the time the solar plasma arrives, the shooter has already escaped from sight, around the western limb of the Sun.

Sources and Characteristics of the Solar Wind

A great deal has been learned about the solar wind and its origins on the Sun since it was first postulated by Eugene Parker in 1958 and then confirmed, by direct measurements in space four years later, in 1962. In the forty-odd years since then, scores of spacecraft have explored and monitored the composition, velocity, embedded magnetic fields, and temporal fluctuations in the flow, employing ever more sophisticated sensors.

To accomplish this, solar wind detectors have sampled and monitored conditions within the heliosphere from well inside the orbit of Mercury to far beyond that of Pluto.

Ulysses is the first spacecraft to explore the solar system above and below the plane of the Earth's orbit: truly the *terra incognita* of the solar system before the itinerant spacecraft was launched in early October, 1990. Since that time *Ulysses* has traced out successive six-year orbits around the Sun, passing over both poles of the star to sample and measure the solar wind in these lesser known regions and to observe from above, the flow of solar plasma from the Sun's polar areas.

With the help of *Ulysses*, *in situ* measurements of the solar wind combined with observations from space of the outer corona—where the solar wind is shaped and directed—have allowed scientists to identify the specific regions in the corona where solar wind streams of various kinds originate and the sources of perturbations in the streams that reach the Earth.

⊙ ⊙ ⊙

A melded view of solar radiation emanating from the transition region of the outer solar atmosphere where local temperatures are in the hundred-thousand degree range, and from the million-degree lower corona, made by combining images of the Sun in the EUV and x-ray regions of the spectrum. Of note is the association of hotter, brighter active regions with closed magnetic loops; and coronal holes and bright points in the transition region with depleted portions in the corona above them.

2003/03/12 13:06

The Sun seen in fine detail in the very short wave, x-ray region of the spectrum. Radiation at these invisible wavelengths originates in the lower corona, where temperatures are measured in millions of degrees Fahrenheit. The hot spots evident as bright features identify regions of concentrated magnetic fields: the origins of solar flares and coronal mass ejections, and the foot-points of coronal streamers which extend beyond what is seen here into the higher corona. Areas of lower electron density that appear as extended dark areas, known as coronal holes, are the loci of open magnetic fields and the origins of high speed streams in the solar wind. Within their boundaries, looking much like the lights of cities seen from space, are highly-concentrated, bright points of coronal x-ray emission.

As noted earlier, two types of solar wind—different in speed, composition and place of origin—blow outward from the Sun.

The so-called slow-speed solar wind has an average speed of about 200 miles per second and is made up of ions common to the upper corona of the Sun, from whence it comes. The sources of the slow wind are coronal streamers, which are associated with strong magnetic regions in the Sun's lower atmosphere. In years of minimal activity they are found only in lower solar latitudes. When the

Sun is more active coronal streamers appear at both low and high latitudes, and there are more of them. Because of this, slow-speed steams in the solar wind are more prevalent in years near the maxima of the 11-year sunspot cycle.

SOLAR WIND SPEEDS COMPARED WITH OTHERS

	MI/SEC	MI/HR
A Speeding Bullet	0.4	1440
Slow Solar Wind, average speed	200	900,000
High Speed Solar Wind Streams	470	1,700,000
Fastest Solar Wind	560	2,025,000
Speed of Light	186,000	670×10^6

The high-speed streams in the solar wind race outward at speeds two to three times faster, and their elemental composition is more representative of the inner corona and transition zone.

Their origins can be traced to extended regions on the Sun where there are few sunspots and other signs of magnetic activity. When seen in x-ray images of the inner corona these "quiet" regions appear as dark vacancies or "holes" in the brighter corona around them: like clearings in an aerial view of a forest, which in this case is made up of closely packed coronal loops; or as bald spots at the poles of the Sun. In the outer corona they appear as regions of very low density, largely devoid of coronal streamers.

Within these extended zones of solar inactivity, magnetic lines of force extend radially outward like blades of tall grass above the surface of the Sun, defining open magnetic field lines that are tied to the star only at their base. The closed field lines which connect regions of opposite magnetic polarity in solar active regions restrict the release of hot plasma from the corona, producing the low-speed streams in the solar wind. In contrast, areas of open magnetic field lines present little opposition, allowing solar plasma in the transition zone and corona to escape the Sun at much higher velocities.

In years around minima of the 11-year sunspot cycle, extensive coronal holes cover the polar caps of the Sun, extending downward in places to lower solar latitudes.

Solar Variability

All solar fluctuations that disturb the Earth can be traced to the effects of the strong solar magnetic fields that thread their way through the photospheric surface and into the middle and outer atmosphere of the star. Indeed, were the Sun to rid itself of all magnetic fields, it would hardly vary at all: leaving but a big light bulb in the sky, slowly burning itself out.

Solar activity is a general term used to describe the nature and extent of solar magnetic fields. The venerable and most common index by which it is described relates to the number of sunspots visible on the disk of the Sun at any time. Since 1848, astronomers around the world have for this purpose employed a universal but arbitrarily-defined index, called the Wolf sunspot number, or more commonly, sunspot number, which endeavors to correct for unavoidable differences in solar telescopes, observing conditions and human observers.

The sunspot number defined in this way is calculated daily, but the most common denomination—the $20 bill of sunspot numbers—is the annually-averaged value. When these are displayed as a time-series graph, a cyclic rise and fall is readily apparent, defining an elastic "cycle" that varies in length from nine to thirteen years, with a mean value of about eleven.

Individual cycles, which by convention run from one sunspot minimum to the next, are identified by number, beginning arbitrarily with the cycle (#1) which extended from 1756—when Thomas Jefferson was a boy—until 1766. The most recently completed cycle, #23, which reached a maximum in 2001and a minimum in 2007, was as always, immediately followed *(The King is dead. Long live the King!)* by cycle #24 which should maximize in about 2013. Through history, the 11-year sunspot cycle has enjoyed an almost hypnotic appeal to professionals and amateurs alike, scrutinized chiefly for with what it might be correlated. Needless to say, its statistical examination is a finely-plowed field.

The nearly random variation in the lengths and amplitudes of cycles, and the existence of periods when for decades the number of sunspots falls to very low levels, tell of an internal mechanism within the Sun that is hardly the precise ticking of a clock. If your electrocardiogram looked anything like the graph of annual sunspot numbers, your doctor would have a very worried look on his or her face.

The underlying 11-year cycle is hardly detectable in weekly or even monthly averages, where it is more than masked by short-term variations of greater magnitude. These are imposed in large part by the Sun's rotation, which continually brings different faces of the Sun into view, and by the birth, evolution and fading away of different active regions.

All of these variations have been known for a long time, as has the revealing fact that sunspots doggedly adhere to fixed rules of placement and magnetic polarity, following the same time-worn patterns and well-worn tracks, from cycle to cycle and century to century.

Sunspots generally appear in pairs, oriented along roughly E-W lines: a leader spot of one magnetic polarity, followed by another, just behind it, of the opposite sign. Sunspots are also concentrated at any time within two bands of solar latitude, one in the northern and one in the southern hemisphere, each of which migrates toward the equator in the course of the 11-year cycle. The magnetic polarities of the leading and the following spot are of opposite sign in the two hemispheres, and this, too, switches on cue, at the start of each new 11-year cycle. The same is true of the two polar regions of the Sun, which reverse polarity with each new 11-year cycle.

Thus the time to complete one full solar magnetic cycle—from one polarity to the other, and back again—is about 22 years: which is fast indeed for an object so large and massive. In comparison, the magnetic field rooted in the molten core of our little solid Earth has reversed its polarity but three times in the last five million years, at highly irregular intervals, and through processes that happen far more slowly and far from synchronously at the two poles. How does the Sun make its switch so fast, and where and how does it happen?

Why the Sun Varies

The answers lie in the fact that the Sun has no long-lasting imbedded field like that of the Earth. What we observe on the surface of the star is a conglomeration

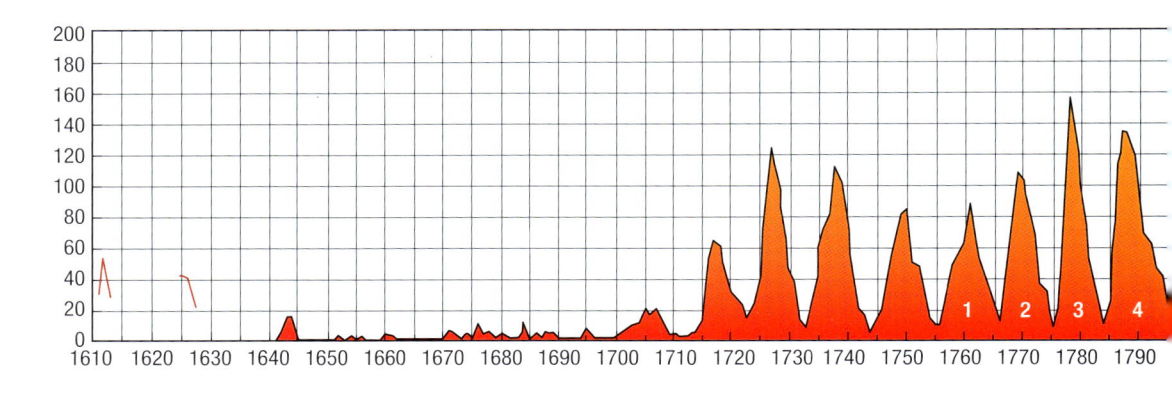

The observed year-to-year variation in the sunspot number (a measure of the number of dark spots and sunspot groups seen on the white-light Sun, corrected for observing conditions) spanning the period from the earliest use of the telescope through 2007. Shown for each year is the mean annual value of daily numbers, which vary considerably from day-to-day.

of superficial fields that are only skin deep: the transitory manifestations of magnetic fields that are continuously generated deeper within the star and carried upward to the radiating surface.

The so-called solar dynamo, which converts polar to toroidal magnetic fields within the Sun, operates on the same basic principle as the huge dynamos in public power plants that convert kinetic energy into electric currents and magnetic fields. But to fit what we know of the ritualistic behavior of sunspots, the solar mechanism must also orient, organize and then release the newly-created magnetic fields in ways that keep them in line and in step, like soldiers on the march, in cadence with non-stop drum beats of eleven and twenty-two years.

Some questions remain, but we now think we know what fuels the solar dynamo, how it operates, and where within the Sun it does its work, initiating a chain of events that ultimately perturbs conditions on our planet and the daily lives of all who live here.

Deep within the hot interior of the Sun, the motions of electrically-charged atomic particles within the solar plasma continually generate incipient magnetic fields, much as the motions of molten metal deep within the Earth give birth to the Earth's magnetic field. But there the similarity ends, for conditions within the Sun and the solar plasma are far more fluid, turbulent and transitory.

When upwardly mobile magnetic fields generated within the Sun arrive at the boundary that separates the spherical, radiative core of the innermost Sun from the deep shell of convection that lies above it they enter a different world.

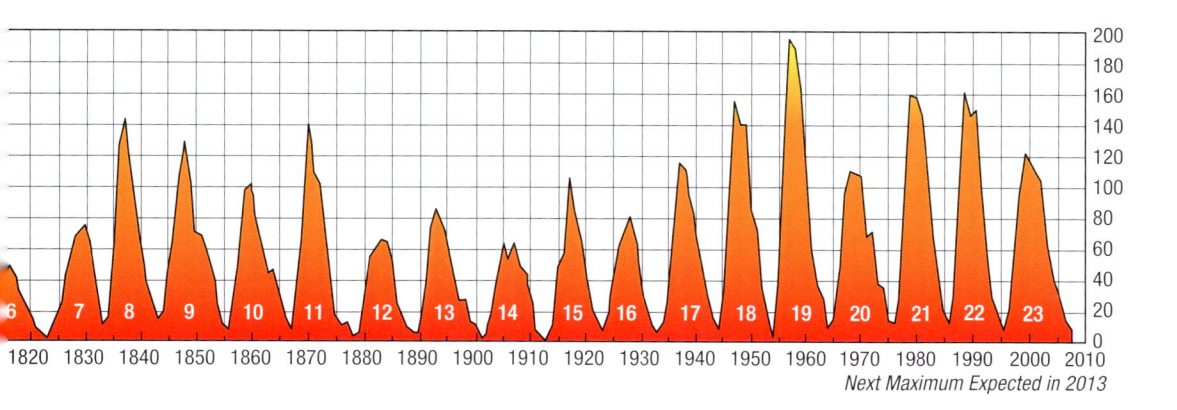

Next Maximum Expected in 2013

Evident is the well-known cycle of about eleven years, and obvious trends in the longer-term, overall level of solar activity. The period of suppressed activity between the mid-1600s and about 1715 is known as the Maunder Minimum, a feature that is also evident in other records of solar behavior.

There in a thin transition layer called the tachocline they come in contact with the strong shearing force that arises from the different rates of rotation of these two internal regions: the one a rigidly rotating ball (the radiative core) and the other an independently-rotating spherical shell (the convection zone) that spins at a different speed just above it. The shearing force at this interface flips and re-orients what were originally N-S oriented fields into those that are organized in the E-W direction, creating closed, magnetic hoop-like rings within the convective zone which lie in planes parallel to the Sun's equator.

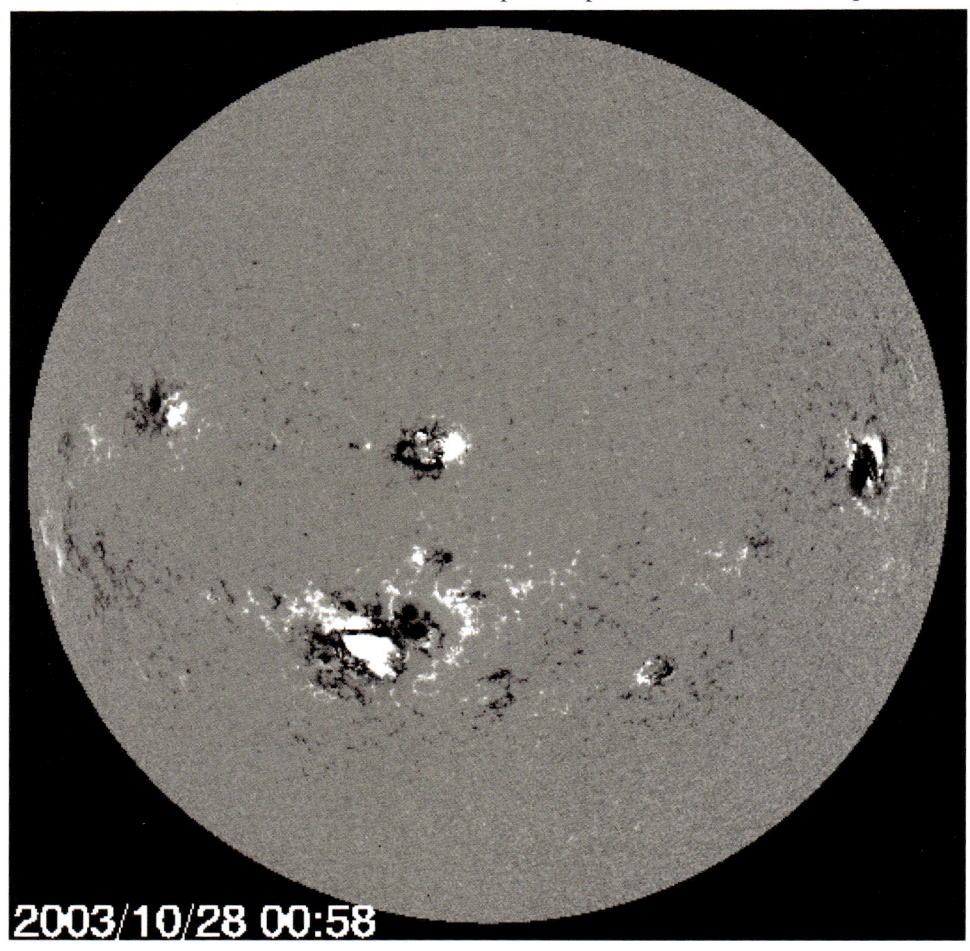

A magnetic picture or magnetogram of the solar photosphere made (from 93 million miles away!) on October 28, 2003 showing regions of strong magnetic polarity. White portrays what is conventionally called *positive* (or *north*) polarity, black *negative* (or *south*) polarity. As we see here, magnetic regions (which correspond to regions of sunspots and other manifestations of concentrated solar activity) are made up of adjacent parts of opposing magnetic polarity which are confined within two distinct belts of solar latitude.

Magnetic pressure pushes these toroidal rings of magnetically-organized plasma upward and outward through the convective zone, to rise like smoke rings, one after another, some into the northern hemisphere of the Sun, others into the southern.

When a portion of any one of these buoyant rising rings of bundled magnetic lines of force comes in contact with the overlying photosphere, the toroid is at that point severed and pulled apart, exposing two ends of opposite magnetic polarity, which emerge as oppositely-polarized regions of surface magnetic activity and most visibly as sunspots.

The number of toroidal magnetic rings released at the base of the convection zone varies in a cycle of about eleven years. Due to the way they were created, the configuration of the magnetic lines of force in rings that approach the photospheric surface in the northern hemisphere of the Sun is opposite in any 11-year cycle to those that take the southern route. The result is a 22-year cycle in the characteristics of magnetic activity in pairs of sunspots in either hemisphere. In one 11-year cycle, as at the poles of the Sun, the polarity of the eastward (or leading) spot of a pair is positive and the westward (following) spot is negative in one hemisphere, and opposite in the other. In the next cycle these conditions are reversed.

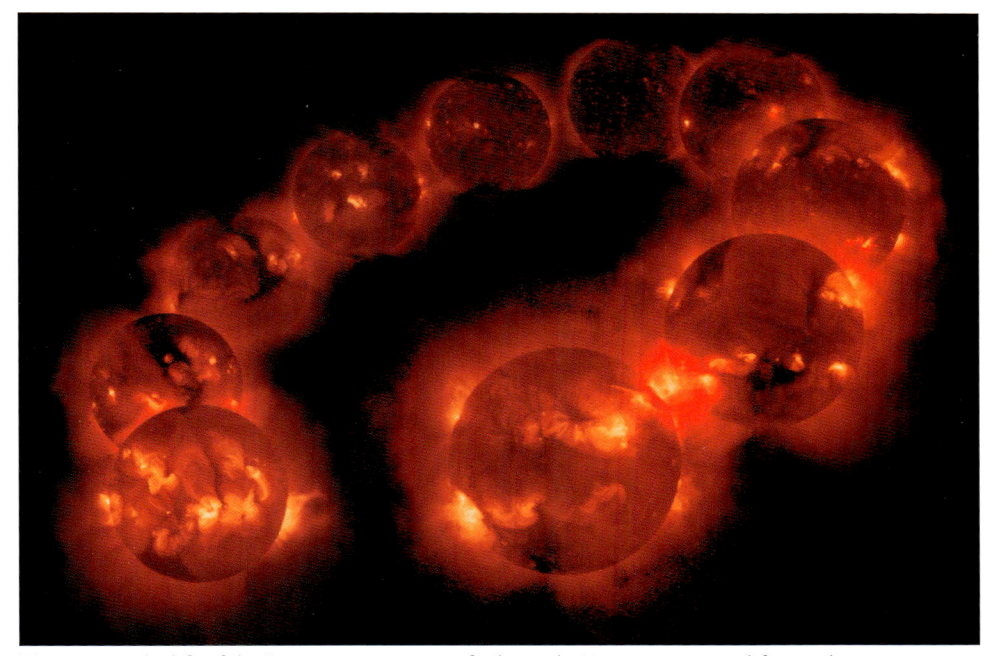

Ten years in the life of the Sun, spanning most of solar cycle 23, as it progressed from solar maximum to minimum conditions and back to maximum (lower left) again, seen as a collage of ten full-disk images of the lower corona made in x-ray radiation. Of note is the prevalence of activity and the relatively few years when our Sun might be described as "quiet".

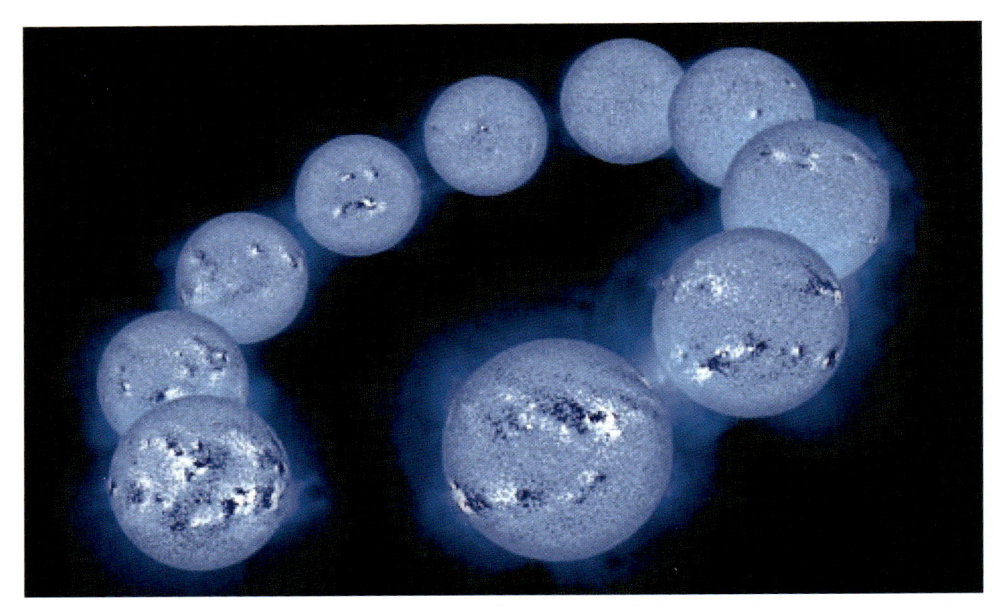

Progress of the Sun through the ten-year period encompassed above, in this case as recorded in corresponding full-disk solar magnetograms that portray magnetic fields of positive and negative magnetic polarity on the surface of the Sun. The two polarities are depicted as either white or blue, with intensity proportionate to magnetic field strength.

Short- and Long-Term Changes in Solar Activity

Recent attempts to do this, based on dynamo models, have ventured testable predictions of when the present 11-year solar cycle (#24) would start and end and how strong it will be. Time will tell, but if proven correct, this could mark a major turning point in our ability to foresee and prepare for many of the societal impacts of year-to-year solar changes.

More difficult to reproduce in dynamo models are the slower and possibly systematic changes, spanning decades to centuries, which are evident in the long record of historically-observed sunspots and in indirectly-obtained proxy data that cover much longer spans of time.

Through the first half of the 20th century, for example, the total number of sunspots observed in successive 11-year cycles steadily increased, in the manner of an amplitude-modulated radio wave, as though the Sun were entering a longer period of higher and higher solar magnetic activity. As well it may have done, peaking, perhaps with cycle #19, as we see in the graph of annual mean sunspot numbers at the bottom of pages 52 and 53.

There have also been extended periods in which the peak amplitudes of successive 11-year cycles were consistently and severely depressed. An example is the period of several decades at the turn of the 18th century, when for several 11-year cycles the number of sunspots dropped to less than half what it was both before and after, and the cycle lengths were unusually irregular. A pronounced drop of a similar amount characterized annual sunspot numbers in three successive decades at the turn of the 20th century.

Most pronounced in the historical, telescopic record of sunspot numbers was the 70-year period from about 1645 to 1715, during which time the total number of sunspots that were observed and recorded was not that much greater than what are seen in a *single year* of high solar activity today.

This latter episode was later named the Maunder Minimum, and one that preceded it, running from about 1450 to 1540, the Spörer Minimum, after E. Walter Maunder and Gustav Spörer, the British and German astronomers who in the late 1880s called attention to the first of these curious anomalies. They were far from the first nor the last to do so, however. The dearth of sunspots between about 1645 and 1715 (which happened to coincide precisely with the reign of Louis XIV, the *Sun King*) had in fact been repeatedly pointed out during the years when it was happening. But it had been largely forgotten and probably discounted following Schwabe's remarkable discovery in 1843 of a 10- or 11-year cycle in annual mean sunspot numbers, which seemed to describe the Sun more nicely as a highly regular and strictly periodic star.

The Maunder Minimum might be considered an artifact in the historical record of telescopically-observed spots on the Sun, were it not that it also appears as a time of dramatic drop in the number of reported aurorae, and as a similar gap in naked-eye sunspots documented in those years by court astronomers of the contemporaneous *Qing* dynasty in China. More important, the Maunder and Spörer Minima stand out as dominant features of the tree-ring record of carbon-14, following seven other similar events—each 50 to 150 years long—that preceded them in time, and as similar features in the beryllium-10 record taken from polar ice cores.

Solar Explosions and Eruptions

Far and away the most dynamic and spectacular changes that occur on the Sun are the short term eruptions and explosions in its outer atmosphere that come and go in but a few minutes, a few hours, or at most a day. These include intensely bright, explosive flares; the eruptions and annihilations of towering solar prominences in the chromosphere and corona; and the expulsion of whole parts of the corona in the form of coronal mass ejections, or CMEs.

The occurrence of these three distinctive types of violent and often-related events is very much affected by the phase and magnitude of the Sun's 11-

year solar of activity. All involve the sudden release or exchange of energy, in massive amounts; and all, not surprisingly, involve the interactions of strong magnetic fields.

In years when the Sun is most active small flares occur somewhere on its surface every minute, and CMEs of one size or another are expelled from the corona at an average rate of three or four per day. In years of minimum activity these numbers fall dramatically.

Explosive Solar Flares

Solar flares appear as sudden and intense brightenings in highly-localized regions on the surface of the Sun: as though a lighted match had been dropped into a puddle of spilled gasoline. The initial and brightest part of the ensuing whoosh of light usually lasts but a few minutes, and the sudden event will run its course, on average, in less than half an hour. Some of the effects on the Earth are almost immediate. Others last for several days.

In the largest solar flares, the amount of energy released on the Sun from a relatively small region and in so short a span of time is far beyond all earthly experience.

⊙   ⊙   ⊙

The asteroid that struck the Earth about 65 million years ago—marking the transition between the Cretaceous and Tertiary periods of geologic history—dealt, without doubt, one of the hardest blows our little planet has ever felt: so momentous that that single event is believed responsible by many scientists for the contemporaneous extinction of the dinosaurs. To produce this extraordinary impact, the asteroid must have been at least five and one-half miles in diameter, as dense as cast iron, and traveling at the moment of impact at about 45,000 miles per hour. Nevertheless, the energy released in that Earth-changing collision was still about 100 times less than what is released on the Sun in a single, very large flare.

As another comparison: The energy released in August of 1945 in the atomic bombs detonated over either Hiroshima or Nagasaki was equivalent to about 20,000 tons of TNT. A single hydrogen bomb—the next step toward destructive awfulness—can unleash the explosive power of *several thousand* atomic bombs. In contrast, the energy released in a large solar flare—of the sort that exploded on the limb of the Sun on Halloween in 2003—was equivalent to several hundred hydrogen bombs detonated all at once at the same place. The internal temperature sustained for a few minutes within the explosion can be for any flare as high as that in the innermost core of the Sun: more than 20 million degrees Fahrenheit. The area on the Sun in which a large flare is concentrated, though a miniscule fraction of the total solar surface, can equal the surface area of half of the Earth, with a vertical extent of about 6000 miles: nearly as high as the Earth is wide.

☉ ☉ ☉

Most solar flares seem to be initiated in magnetic loops of the lower corona, and most of the immediate, ensuing action takes place above the photosphere in the chromosphere and transition zone. In these two intervening regions of the solar atmosphere the sudden and overwhelming impulse of energy—downward from the corona—triggers immediate responses in the form of bursts of radiation and particle acceleration.

Major flares occur only in magnetically-active regions on the Sun: in the distinctive and often snarled and twisted magnetic fields of complex sunspot groups. Their precise points of origin are most often found in places of magnetic confrontation and conflict where oppositely-directed field lines come into contact with each other. When this happens, the magnetic field lines can be severed, reconnected or re-formed, through processes that convert some of their huge store of magnetic energy into heat, radiation, and the acceleration of atomic particles in the local plasma.

The tremendous amount of radiative energy that flares emit is distributed over the full breadth of the electromagnetic spectrum: from gamma-rays to x-rays to ultraviolet to visible to infrared and radio waves. When a large flare occurs on the Sun it can be detected by instruments ranging from x-ray telescopes to radio receivers.

The high-energy x-rays, gamma rays, and far ultraviolet radiation from this blast of converted energy are sprayed out from the source of the flare in all directions, and strike the upper atmosphere of the Earth with no warning at all, since they too, travel at the speed of light. The heavier atomic particles that receive the most energy can be driven from the Sun at velocities of up to almost half the speed of light, to arrive at the Earth in but fifteen minutes, which is but seven minutes after the flare is first seen.

Slower particles of lower energy will strike the Earth within hours to a few days later, but they can affect the Earth in only minor ways, compared to the potential damage done by the more energetic solar protons in the initial eruption or the effects of CMEs. Radiation and heavy particles from large flares pose a direct hazard to exposed spacecraft and astronauts whose travels take them beyond the protective shields of the magnetosphere and upper atmosphere.

Three images showing the eruption of an extremely energetic solar flare at the limb of the Sun, taken at the same time on November 4, 2003. Shown, from bottom to top, are the white-light photosphere; the upper chromosphere (seen in the extreme ultra-violet); and in x-ray wavelengths, the lower corona (from which most of the flare's prodigious store of energy was released.) The associated magnetic region is visible in the first of these as a cluster of large sunspots being carried over the east edge of the Sun by solar rotation. The actual size of the associated active region is revealed in the two higher images as an extensive group of very large magnetically-active regions.

Solar Prominences and Filaments

Some of the magnetic lines of force that reach upward from regions of particularly strong sunspot magnetic fields are able to carry cooler chromospheric material with them into the hotter corona, where it is magnetically insulated and held aloft for a time, in the form of extended clouds called solar filaments.

When seen in visible light at the edge, or limb of the solar disk, as in a coronagraph or when the Sun is eclipsed by the Moon, these extensions of

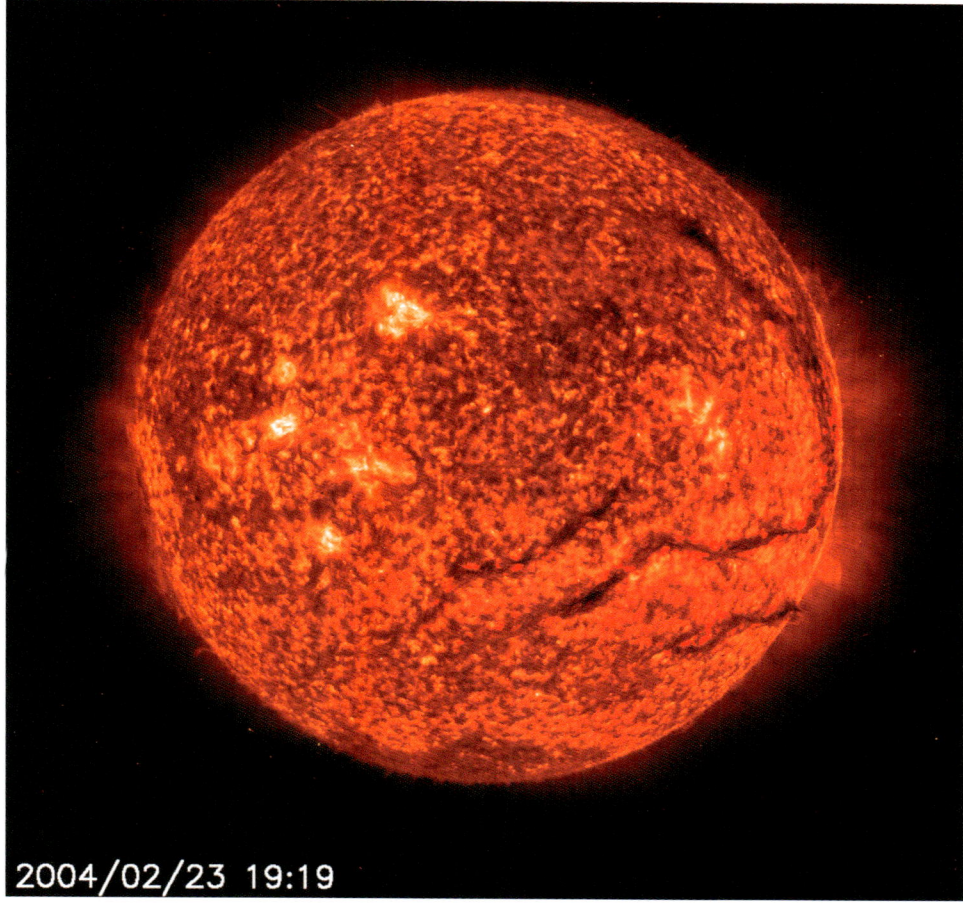

2004/02/23 19:19

The active Sun seen in the EUV on February 23, 2004, displaying hotter, magnetically-active regions and cooler, darker filaments. Dark filaments of this kind, were they seen at the edge of the Sun, would appear as long, elevated clouds extending above and beyond its curved limb. They can persist for several weeks and are composed of hot chromospheric material held aloft in the less dense and far hotter corona by arched magnetic field lines, defying the force of solar gravity and the fundamental principle of heat transfer. Were it straightened out, the sickle-shaped, longest of these would cover nearly four times the distance that separates the Earth from the Moon.

2005/07/28 22:00

Solar rotation carries an active region into view on the east limb of the Sun, revealing the initial lift-off of a newly-spawned coronal mass ejection. In time, as it speeds outward into interplanetary space, the CME will expand to a size larger than the Sun itself. This view of the chromosphere, made in the far ultraviolet light of ionized helium, shows the solar plasma at a temperature of about 100,000° Fahrenheit, ten times hotter than the spotted, white-light photosphere which lies just beneath it.

chromospheric material, aptly called solar protuberances or prominences, appear in varied sizes and graceful shapes—sometimes resembling the handle of a Sun-sized teacup. Under these circumstances they stand out against the dim white of the corona as brighter features which are distinctively red: the pure color of light emitted by excited hydrogen atoms.

When we look at the disk of the Sun in the red light of hydrogen in the chromosphere, the same prominences—now seen from above—appear not brighter but *darker* than the surrounding disk, as worm-like dark filaments that are both cooler and denser than most of the other material around them.

When seen from this perspective it is obvious that the often sinuous forms of solar filaments follow the course of the neutral dividing line that separates regions of opposite magnetic polarity in solar active regions: like a referee who with arms extended, attempts to hold opposing magnetic pugilists apart.

When the attempt fails and the two oppositely-charged regions come in contact, the disruption and reconnection of the magnetic lines of force that had suspended and nourished the great mass of chromospheric plasma are suddenly unleashed, with the same force and some of the same effects as an explosive solar flare. When this happens, the filament can be torn loose from its magnetic roots at the solar surface and flung outward, bodily, from the Sun.

A rising solar filament, driven upward and outward at speeds of several hundred to a thousand miles per second is but one manifestation of a more extensive magnetic disturbance in the corona which expels much larger loops of coronal plasma at far greater speeds. In the course of these larger-scale magnetic reconfigurations, entire streamers can be torn from the corona and thrown outward into space, like petals pulled from a daisy. These immense coronal mass ejections (or CMEs) expand as they proceed outward from the Sun, and soon exceed by far the size of the star that gave them birth and sent them on their way.

Filaments or prominences are common features on the Sun at times of increased solar activity. Those known as active prominences, which straddle a region of magnetic activity with feet in regions of opposite magnetic polarity, can erupt within a few days of their formation. Another class, found in quieter and less rowdy places on the surface of the Sun, can literally hang around for months at a time, peacefully and with little change, until they slowly fade away.

They can also grow into immense structures that seem wholly out of scale when compared with anything else on the surface of the Sun: truly gargantuan features when we see them extended outward above the limb. Or on the disk, where as dark filaments they look a lot like segments of the Great Wall that snakes its way across the northern hills of China.

The largest of all are the long-lived and relatively inactive quiescent prominences that can tower tens of thousands of miles above the surface of the Sun, and extend for hundreds of thousands of miles. Were one of these large quiescent prominences somehow suspended in the space between the Earth and the Moon, it would span most of the distance that keeps us apart: a red bridge of many arches in the sky, nearly 200 thousand miles long.

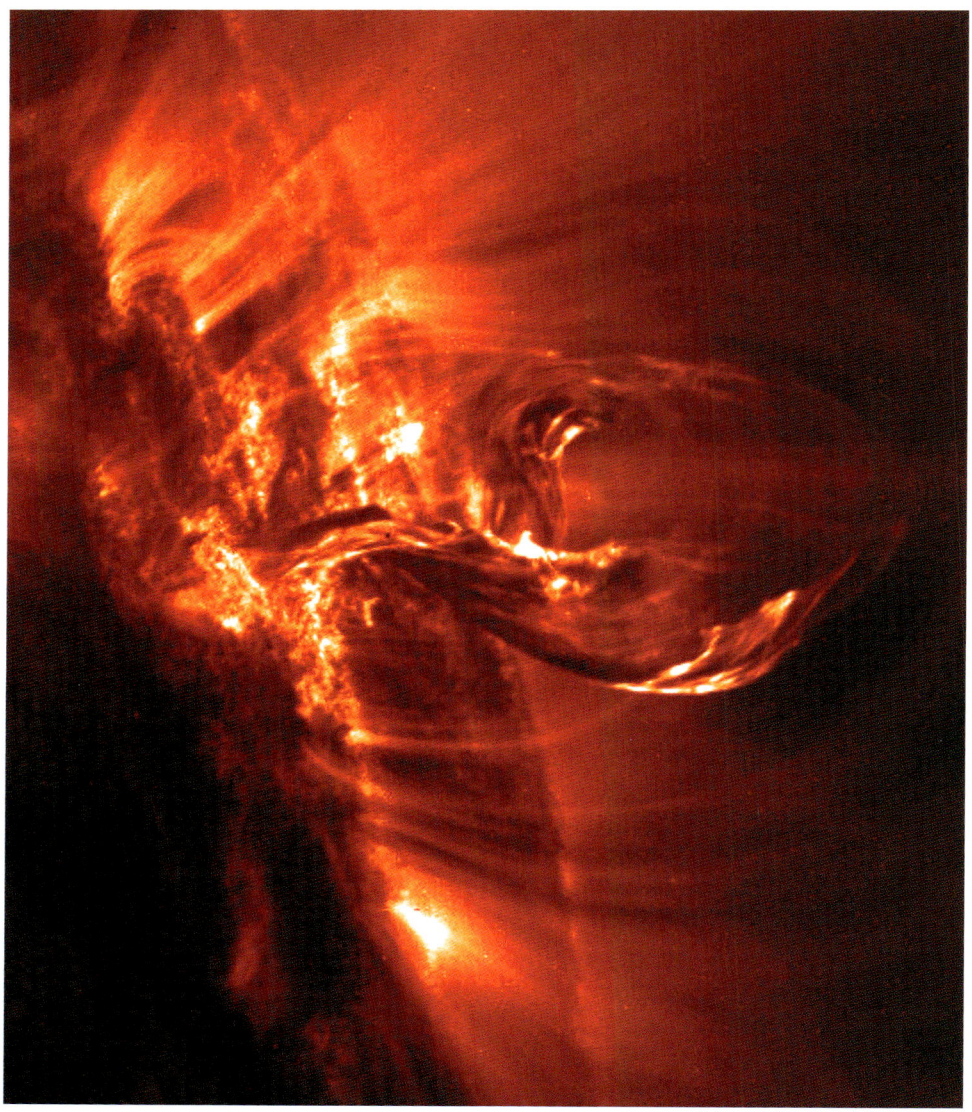

The eruption of a looped solar filament that is rooted in a magnetically-active region near the apparent edge, or limb, of the Sun. The image, from the TRACE spacecraft, was made in the light of the (invisible) extreme ultraviolet spectrum, emitted from regions of the solar atmosphere where temperatures exceed more than two million degrees Fahrenheit.

Coronal Mass Ejections

If sunspot magnetic fields are the gunpowder, flares the muskets and prominences the horse-drawn cannons in the venerable solar armory, coronal mass ejections or CMEs—which came to be recognized but thirty years ago— are truly the heavy artillery. Indeed, interplanetary CMEs are the primary drivers of almost all space weather disruptions, including highly accelerated plasma streams and most major geomagnetic storms, with potential impacts on a wide range of human activities.

CMEs are the product of impulsive changes in the corona, in which entire segments of the outer corona are driven outward from the Sun and ejected into the solar wind stream. They originate in regions of closed magnetic field, most often from the coronal streamers that extend outward from the disk of the Sun like the petals of a flower.

The bulbous, helmet-shaped base of a coronal streamer is formed by closed magnetic fields that are strong enough to contain the coronal plasma at this level and keep it from expanding and escaping as solar wind. This is not the case, however, for the outer extensions of streamers. There the strength of the coronal magnetic field, which decreases with height, can no longer contain the coronal expansion. At this level—between about 0.5 and 1 solar radius above the limb—the coronal plasma is able to stream freely outward from the Sun, producing the typical tapered form of extended coronal streamers.

It is largely the inner parts of a streamer, however, that are expelled from the Sun as CMEs. When the magnetic bindings that contain the base of a coronal streamer are disturbed or catastrophically broken, the plasma that was confined within it—often an entire coronal streamer—is flung outward as from a loaded spring. As it moves outward, the outer edges of the expanding CME often appear as a closed loop, still attached at the Sun, and indeed, few CMEs ever completely sever their magnetic connection to the star.

There is no general agreement as to what initiates the release of a CME, although stressed magnetic fields are undoubtedly involved. One possible explanation holds that when the two foot-points of a large coronal magnetic loop—one rooted in a magnetic region of positive polarity and the other in the opposite sign—are moved relative to each other by the shearing action of differential rotation at the surface of the Sun, the towering loops that form the base of an overlying coronal streamer are directly affected. A likely result is a twisting of the loop that can be relieved only by dynamic realignment and readjustment, with the explosive release of some of its vast store of magnetic energy.

It is also clear that CMEs are a significant player in the evolution of the corona from minimum to maximum levels of solar activity, and that the release of a CME is quite likely a stress reliever that enables the outer solar atmosphere to reconfigure itself in response to underlying changes in the solar magnetic field.

CMEs are expelled from the Sun at a wide range of speeds: the slowest cover a few tens of miles in a second, and the fastest move a hundred times faster. At these speeds slower CMEs will travel a distance equal to the radius of the Sun (430,000 miles) in but a few hours, and the fastest in but a few seconds. It is rare that one can distinguish the movement of anything on the Sun as it happens because the Sun is so far away. But CMEs are so large and fast-moving that through a coronagraph—where most of them are detected and tracked—one can follow the initial outward progress and expansion of many of them in real time.

Coronal streamers that give rise to CMEs often become more luminous a day or so before the eruption, as though signaling their own impending demise. And in time, moreover, a replacement streamer may re-form in the vacant space where the original was torn away. What has happened, in any case, is a violent restructuring of a large part of the solar corona, an expulsion of a piece of it which expands like a balloon as it moves outward, and the release of a large amount of energy into interplanetary space.

Carried off with the expanding plasma cloud is the embedded magnetic field that gave it form and structure. Any prominence that lies within the bulbous base of the same streamer is also often torn away. These are commonly uprooted and carried away from the Sun with the CME, initially as a cooler and denser arch of chromospheric material enclosed within the expanding coronal loop, though driven outward at a somewhat slower speed.

Often solar flares are also seen in the same active region on the Sun, at about the same time or shortly after the eruption of a CME. But the flares and eruptive prominences that are so often associated with coronal mass ejections do not necessarily provoke the CME, or the other way around. They are, rather, different consequences of the same large-scale, magnetic event.

☉ ☉ ☉

Although some move faster, the average speed of a CME through the outer corona—about 250 miles per second—is still far short of the velocity needed in the corona to overcome the immense restraining pull of solar gravity. Yet they escape. This and the fact that none of them fall back to the solar surface—in the manner of arrows shot into the air—tell us that other forces within the corona continue to accelerate a CME, once launched, to send it ever faster on its way.

CMEs could be compared to tornadoes, hurricanes and tsunamis on the Earth were these coronal expulsions less frequent and commonplace. As noted earlier,

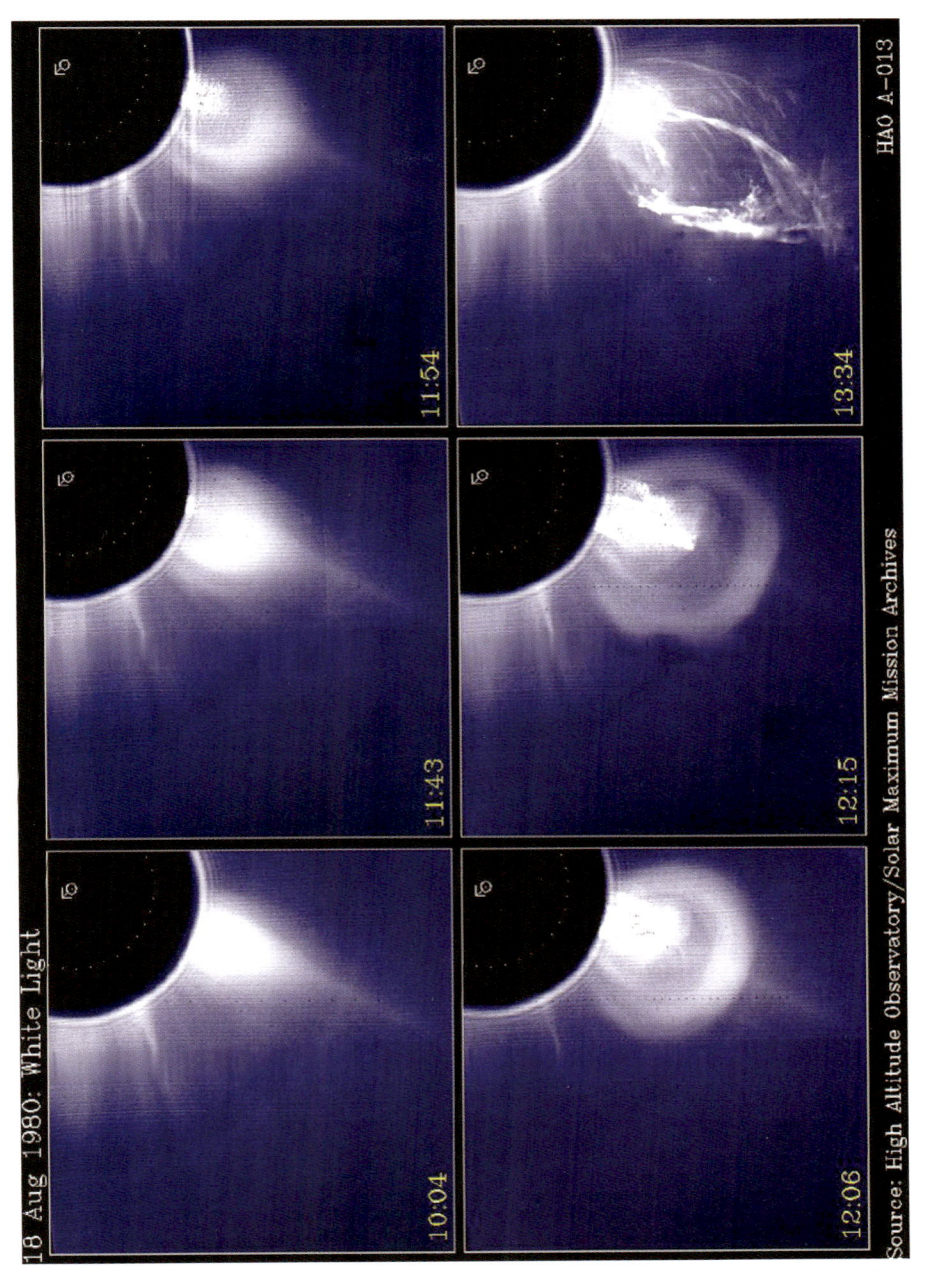

The formation and release of a mass ejection (CME) in August of 1980, near the maximum of solar cycle #21, showing its destructive impact on the coronal streamer of which it was a part. The sequence of six images (arranged from bottom left to top left, then bottom right to top right, cover a period of 3½ hours, with most of the major changes in the half hour captured in images 2, 3, 4 and 5, which are each about ten minutes apart. The loop-like CME seen in the last image, with a mass of about ten million tons, will speed outward from the Sun into interplanetary space at a million miles per hour or more. CME's were not discovered until the early 1970s, when solar coronagraphs capable of seeing them were put into space.

how often they occur generally follows the phase of the 11-year sunspot cycle. On average there are between three and four CMEs per day when the Sun is near the peak of its activity cycle. At times of minimum activity, the number falls to about one every ten days. Thus, when averaged over the entire cycle, the Sun produces on average of about one per day, like a hen laying (very large) eggs.

In the process of this continuing celestial workout, the Sun sheds a lot of pounds, whether reckoned per month, per week, or per day; and unlike us, none of the lost weight will ever be regained. A typical CME carries off a thousand million, i.e., a billion tons of coronal plasma. In one month, at the average rate of one CME per day, the Sun loses an amount of mass equivalent to the weight of all the water in the five Great Lakes, one of which is the largest fresh water lake in the world.

The expulsion of a CME from the Sun, made up of an entire coronal streamer and the chromospheric prominence that lay within it. The diffuse remains of the streamer (at about 11 o'clock) is, like the rest of the white-light corona, composed of solar electrons illuminated by sunlight from the here-hidden photosphere. The ejected prominence that trails behind and slightly to the left of the coronal mass retains its original looped shape, though now in greatly expanded form. The size of the solar disk, dwarfed by what it has sent into space, is shown as a white circle in the center of the larger circular region of the inner corona that is blocked by the coronagraph.

But the corona isn't made of water, and the Sun is not the Earth but an object that is 330 thousand times more massive. Had the Sun sent out not one but a hundred CMEs each day since it first became a star, 4.6 billion years ago, its total mass would by now have been reduced by about .01%: a fractional change which for a 210 lb. dieter would correspond to a total weight loss (after all this time and all that effort!!) of less than a third of an ounce.

PRINCIPAL MANIFESTATIONS OF SOLAR ACTIVITY

FEATURE	DESCRIPTION	LIFETIME	EFFECT ON THE EARTH
Sunspots	Dark spots that appear on the white-light disk of the Sun, each the locus of strong and concentrated magnetic flux	Days to a month or more	Incremental reduction in the total solar radiation received at the Earth
Plages	Bright patches in the photosphere and chromosphere, most often in association with sunspots	Days to a month or more	Incremental increase in the total solar radiation received at the Earth
Flares	Sudden, explosive brightenings in magnetically-active regions in the chromosphere and corona, accompanied by the release of highly energetic electromagnetic radiation and atomic particles	A few minutes to an hour or more	Drastic, transitory increase in the flux of x-ray and extreme-ultraviolet radiation with effects on the upper atmosphere; release of highly energetic protons, posing hazards to manned and unmanned spacecraft and jet aircraft at high latitudes
CMEs	Expulsions of large segments of the outer corona, often becoming larger than the Sun itself, into inter-planetary space	Days to weeks	Acceleration of atomic particles in the solar wind, with concomitant impacts on the magnetosphere and upper atmosphere

An artist's depiction of the environment of the spherical Earth, at center, seen against a background of other stars, as it would appear from space were we granted the gift of magnetic vision. The source of the nearly symmetric pattern of magnetic lines of force attached at the Earth is the internal magnet that exists within the planet's dynamic metallic inner core. Several thousand miles above the buried magnet, lines of force of opposite polarity emerge from regions near the Earth's rotational poles. They trace out the familiar pattern of a simple bar magnet, distorted by the force of the solar wind which compresses it on the Sun-facing (upper left) side. Trapped within the confines of the inner, stronger lines of magnetic force are charged atomic particles—electrons and protons—which are concentrated in two concentric radiation belts, the innermost of which is portrayed here in cross section, like a sliced doughnut.

THE NEAR-EARTH ENVIRONMENT

A Protected Planet

We live, by chance or otherwise, on a highly protected planet. Were it not for the two natural barriers that stand between the Sun and the surface of the Earth, life itself would soon disappear.

These essential shields—the magnetosphere and just beneath it, the gaseous atmosphere of the Earth—protect us from the full fury of the highly variable star with which we live, and from even more energetic atomic particles that arrive from distant cosmic explosions.

The first of these, the magnetosphere, is an invisible cage built of magnetic lines of force that deflect the vast majority but not all of the high energy electrons, protons, and ions that continually assault the Earth on their way outward from the Sun. But this over-arching shell offers no defense at all against a second and equally persistent hazard: the onrushing streams of highly energetic ultraviolet, x-ray and gamma ray photons that pour outward from the Sun and against the Earth, day after day.

Atoms and molecules of oxygen and nitrogen in the air provide this second shield by filtering the electromagnetic radiation that enters the top of the atmosphere. In this wavelength-selective process, solar heat and visible radiation are allowed to pass through the atmosphere from top to bottom to warm and illuminate the surface of the Earth, while solar gamma rays, x-rays, and the most damaging ultraviolet rays are absorbed and thus removed.

The same atoms and molecules of air also serve as a second line of defense against high energy atomic particles, including those of neutral charge.

The incoming solar or cosmic particles that make their way through the magnetosphere and into the atmosphere lose much of their energy in the course of collisions with atoms and molecules of air, and in the process are de-fused: just as the same all-protective blanket of air shields the surface of the planet from the force of all but the largest meteors that enter the top of the atmosphere. It also keeps us comfortable by holding in much of the heat released from the Sun-warmed surface of the Earth.

Beyond the outer edges of these two trusted bulwarks—the atmosphere and the magnetosphere—lies a world that is far more hazardous: the airless, lifeless heliosphere, or realm of the Sun, that reaches outward from the star in all directions for ten billion miles and more.

The Air Above Us

Ulf Merbold, a German astronaut, may have best described the Earth's atmosphere when late in 1983, from the vantage point of space, he first looked out at it from above, through a window in the Columbia space shuttle:

> *"For the first time in my life I saw the horizon as a curved line. It was accentuated by a thin seam of dark blue light—our atmosphere. Obviously, this was not the ocean of air I had been told it was so many times… I was terrified by its fragile appearance."*

The blue blanket of air that warms and protects the planet is indeed fragile and surprisingly thin, given all that it must do. The atmosphere of the Earth extends above the surface in the form of air to a nominal height of roughly 100 miles, but not as a uniform or homogeneous layer. More than half of the air that it contains is found within the first ten miles of that distance, and when one is less than a third of the way to the top so much of the molecular air is gone that the daytime sky is no longer blue, but black.

Moreover, along the way, with ever-increasing height above the ground, the atmosphere undergoes major transformations: not only in density and temperature, but in composition and function as well. Were it a country, the atmosphere would be described by geographers and demographers as a loose federation of vertically-separated ethnic regions, each quite different in population density, demographic characteristics, prevailing climate, and the work that gets done there.

Changes on the Way to the Top

Obvious to anyone who has ascended a mountain is the change in the density of the atmosphere with altitude: how much lighter, thinner and headache-ier the air becomes the higher up we go.

In Denver, but a mile above the sea, the density of the air—and hence the amount of oxygen we take in with each deep breath—has already dropped by 15% when compared with air in Boston or San Francisco. On top of Pike's Peak, at 14,100 ft, the oxygen in each breath has been reduced by more than a third from sea-level values, and atop Mt. Everest—at 29,035 ft, the highest point on the planet—by more than half.

The flash of the rising Sun as it appeared in space, above the darkened night-side of the Earth, illuminating the thin, protective shell of our atmosphere (curved blue line). At lower left we see the first glint of the Sun on the vertical stabilizer and other parts of the Space Shuttle Discovery, from which this picture was made. Due to the speed of its orbit around the Earth, astronauts witness a sunrise and sunset every 90 minutes.

So much of the Earth's atmosphere is so vacuous, diaphanous and diffuse that were we to compress its 100 miles or so of thickness into a layer of uniform density, like the air near the surface, we would be left with a sky but five miles high with no air to breathe above it. In squeezing the atmosphere down in this way, the peaks of at least a dozen Himalayan mountains—no longer covered with air—would protrude above it like a chain of islands in mid-ocean.

The *temperature* of the ever-thinning atmosphere also changes with altitude, but neither uniformly nor always in the same direction. A concentrated layer of ozone in the middle atmosphere, created there by the effects of incoming solar ultraviolet radiation on molecules of oxygen, alters what would be a much simpler profile of temperature vs height.

Were the ozone layer not there we would expect the temperature of the air to cool with height above the warm surface of the planet. At some level part way to the top, as the air becomes thinner and thinner, the cooling trend would slow and reverse direction, now heating up with height, due to the absorption of solar short-wave radiation in the upper atmosphere. In that simplest case the profile of temperature versus height above the surface of the Earth would trace out a smooth curve that looked like a big "C" when plotted with warmer temperatures to the right and colder temperatures to the left.

But the stratospheric ozone layer alters this simplified relationship by adding heat to the middle atmosphere, which interrupts the expected cooling with height at an altitude of about ten miles above the surface and reverses its direction. What is then a *warming* with height continues through a run of another 20 miles in altitude, where, at the end of the ozone layer, the atmosphere reverts once more to its initial trend of cooling with height. Much higher, as we approach the top of the atmosphere the cooling trend slows to a stop and reverses itself one more time, now heating monotonically with height, in this case driven by the absorption of solar EUV and x-ray radiation.

With these forced reversals from cooling-with-height to warming, to cooling, and back to warming again the expected "C" shape of the temperature profile of the atmosphere (with a single temperature minimum about one-third of the way to the top) is bent into a more complex curve (shown in red on the following image), with minima at about fifteen and eighty miles above the surface and a maximum in between. These diversions and excursions divide the atmosphere, in terms of temperature and function—into five distinct horizontal layers: the troposphere, stratosphere, mesosphere, thermosphere, and exosphere.

⊙   ⊙   ⊙

But in speaking of the "temperature" of the atmosphere we refer not to the effective temperature which you and I would feel were we to ascend in a balloon gondola through these higher regions of the atmosphere, but the physically-defined kinetic temperature: a measure of the speed of random motions of atoms and molecules of air, no matter how thin or densely packed they are.

How much a person would feel these random movements of the sub-microscopic constituents of air—moving slower at lower temperatures, faster at higher—depends almost entirely upon the density of the air, which in the Earth's atmosphere decreases markedly with altitude. Thus an astronaut

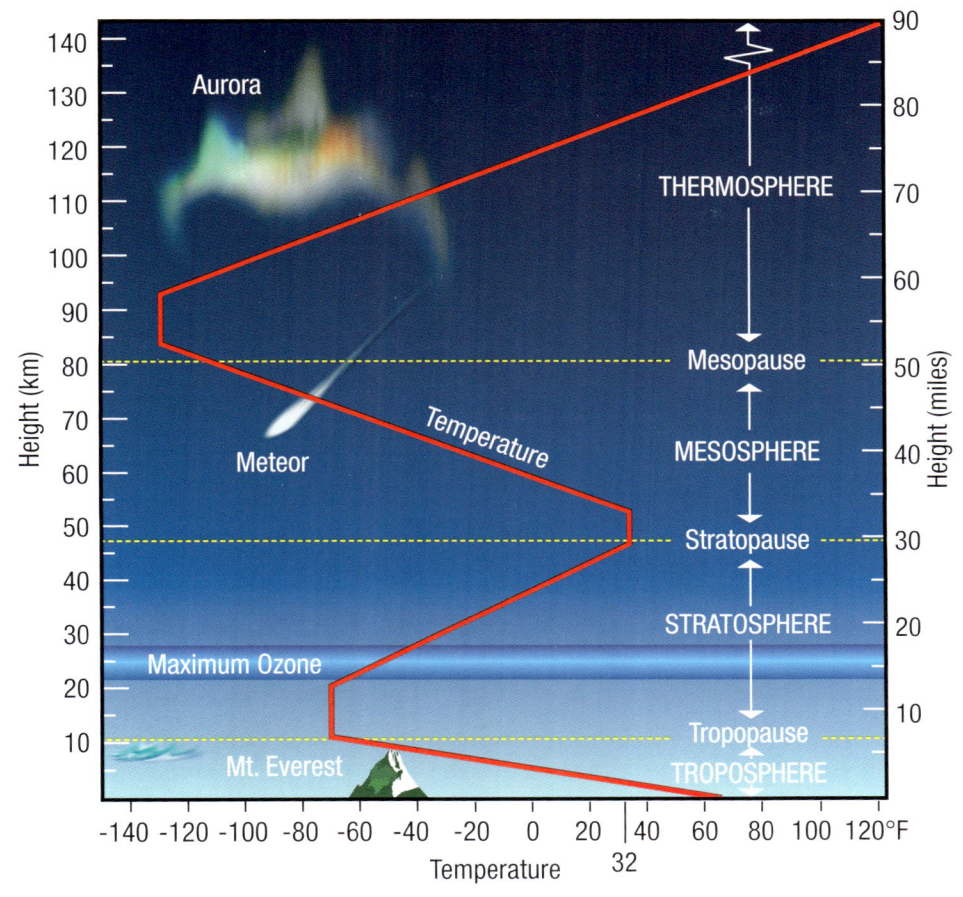

The average temperature of the air above our heads, shown as a red line, as it varies from the surface of the Earth to a height of 90 miles (about 145 kilometers.) Also shown are the conventional horizontal layers of the atmosphere, which are defined by systematic changes in temperature; Mt. Everest, at 29,028 feet, the highest point on the surface of the Earth; the maximum concentration of ozone in the lower stratosphere; and the upper-atmospheric region in which energetic particles from the Sun or magnetosphere excite atoms of air to produce aurorae. Incoming solar particles are chiefly blocked or attenuated by atoms and molecules of air in the thermosphere, although some secondary neutrons from cosmic rays make it all the way down into the troposphere.

working outside the space station in the thermosphere—some 250 miles above the surface of the Earth where the kinetic temperature can range from 700 to more than +1000°F—would be immersed in so rarified a medium that to him or her it would feel very cold.

The Troposphere

The lowest of these five layers of the atmosphere is our home, the troposphere. It is also the thinnest layer, extending upward to a height of but about seven miles, or about 40,000 ft: roughly half again as high as Mt. Everest. Were this layer of mostly breathable air included as a transparent shell on a 12-inch library globe it would be about as thick as the paper of this page. Yet within this thin veneer is almost all life on Earth, and the stage on which all but the last few years of human history have been played out.

Within the troposphere are almost all of the moisture in the atmosphere, all weather and climate as we know them, and all but the rarest of clouds. The greenhouse gases—principally water vapor, carbon dioxide and methane—that trap escaping terrestrial heat to keep the planet warm are found mostly within the troposphere.

Unlike all the rest of the atmosphere, the troposphere is an integral part of the biosphere, in that the air in this layer is connected interactively with the oceans, the land, and living things through biogeochemical exchanges of carbon, nitrogen, oxygen and other life-essential elements.

The tropo in tropo-sphere was chosen to indicate "turning" or "change," and more specifically, the steady drop in temperature that distinguishes this lowest layer of the atmosphere as one ascends higher and higher above the solar heated surface of the Earth. In seven miles (somewhat more at the equator and less at the poles), the temperature falls through about 130°, from a global annual mean value of about 60° F at the surface to 70° below zero at the very top, where the next layer, the overlying stratosphere, begins.

The Stratosphere

At the transition between the thin troposphere and the more extensive stratosphere—called the tropo*pause*, meaning *end of the troposphere*—the cooling trend reverses direction to follow a smooth and continued warming that persists through the next 25 miles in altitude.

At one time the temperature of the air above the troposphere was thought to remain more nearly constant with altitude, and hence the name *strat*, meaning *stretched-out* or *extended* atmosphere. In fact, in the course of the climb through the stratosphere, the air heats up by almost 100° F. There, about 32 miles above the surface of the Earth, the temperature of the air (now about +20° F) is back within the range of wintertime temperatures on the ground, which, were the air more dense, would be comfortable enough to enjoy in a heavy coat, scarf and stocking cap.

The lowest stratosphere is the region where most commercial jet aircraft fly, at altitudes of six to eight miles above the surface. The instrumented meteorological sounding balloons that are routinely launched, by hand, every day around the world, make their measurements from ground level to an altitude of about 22 miles, more than halfway to the top of the stratosphere.

This distinctive 25-mile layer of heated air exists because within its bounds is almost all of the ozone in the atmosphere.

Molecular ozone is a voracious but highly selective absorber that removes almost all of the incoming solar ultraviolet radiation that penetrates to this depth in the atmosphere, and in the process, warms the air around it. In performing this function the ozone layer—which reaches down to a scant 8 miles above our heads—stands as the last line of defense in the atmospheric shield: to protect life on the planet from the daytime dose of potential lethal ultraviolet radiation from the Sun.

The intimate relationship between ozone and near-ultraviolet radiation is a give-and-take affair. Solar radiation in the near-ultraviolet region of the spectrum *produces* ozone by breaking down molecules of ordinary oxygen in the stratosphere. But the ozone created there by the Sun in turn *absorbs* some of the incoming solar ultraviolet radiation in an adjoining region of the ultraviolet spectrum.

Through a balance of these opposing effects, changes in the amount of ultraviolet energy we receive from the Sun, as in the course of the 11-year solar activity cycle, alter the amount of ozone in the stratosphere, as do the high-energy solar particles that come our way in the course of solar explosions and eruptions. And in these modern times, our own activities—miles below the stratosphere—have clearly reduced the amount of ozone in the ozone layer, through our release of ozone-destroying chemicals, industrially, agriculturally, and personally, at the surface of the Earth.

Our gaseous atmosphere is composed mostly of nitrogen and oxygen in molecular and atomic form: ozone, a "trace" constituent, accounts for less than one part per million, by weight or volume. Moreover, a protective ozone layer is found on no other planet in the solar system, and the little we have is endangered by what we do.

The Mesosphere and Thermosphere

At the stratopause, or *end of the stratosphere*, the scant ozone that still remains can no longer heat the atmosphere, and there—a little more than thirty miles above the surface—the temperature of the air again begins to drop, resuming the kind of cooling with increased altitude that was the hallmark of the troposphere. The cooling trend, though not as steep as that near the ground, persists through the next 22 miles, through the overlying *meso*, or *intermediate* atmosphere: called the mesosphere. Where the cooling ends—a little more than 50 miles above the surface, at the mesopause—the air temperature has fallen to minus 135° F, the lowest temperature found anywhere on Earth or in its atmosphere.

Molecules of oxygen and nitrogen in the mesosphere absorb a portion of the incoming solar ultraviolet radiation and as a result are broken apart into single atoms of these gases. Beginning about halfway up through the mesosphere, at a little more than 40 miles above the surface, these gases are present in the atmosphere in both molecular and atomic form, with the balance shifting with altitude toward the latter.

Ten miles higher, just above the mesopause—in the thermosphere, or *heated atmosphere*—the profile of air temperature versus height undergoes its third and final reversal, and the air once more begins to warm with increasing altitude. The source of the added heat, as in the stratosphere, is again the absorption of short wave radiation from the Sun, but this time in the more energetic extreme-ultraviolet and x-ray region of the solar spectrum. The heat absorbers in the thermosphere—a vast region that stretches from 50 to more than 600 miles above our heads and far above the "nominal" height of the atmosphere—are the last remaining atoms and molecules of both oxygen and nitrogen, which at these lofty heights provide the first line of defense against the full blast of incoming solar short-wave radiation.

☉ ☉ ☉

The end effect, in the lower half of the thermosphere, is a steep and phenomenal rise in the temperature of the air, becoming ever hotter with altitude due to the increased exposure of the air to the direct short-wave radiation from the Sun. From a starting value of 135° below zero at the base of the thermosphere—about 50 miles above the surface of the Earth—the air warms so steadily with altitude that but 20 miles higher, temperatures in the solar-heated thermosphere are in the range of those experienced in southern Arizona in the summer. Just ten miles higher—some 80 miles above the surface of the Earth—the temperature of the air has reached the boiling point of water, although the air is far too thin to do it.

And on and on it warms, without respite until—at an altitude of about 300 miles—these kinetic air temperatures in the sun-lit thermosphere have flattened out in the range of 900 to 2000° F, depending on the highly variable amount of x-ray and ultraviolet radiation that the Sun happens to be releasing at the time.

The tenuous air in the thermosphere, moreover, has little thermal inertia, or heat-holding power, and as a consequence, the air temperature goes through extreme bimodal swings from night to day, as well as strong seasonal variations, and dramatic changes in the course of changes in solar activity.

☉ ☉ ☉

The thermosphere is the region of the atmosphere where many Earth-orbiting spacecraft fly, and the place where the meteors that find their way into the atmosphere are heated by friction to produce shooting stars and meteor showers. It is also the site of auroral displays: the northern and southern lights which are most often seen at higher latitudes in the nighttime sky.

The Ionized Upper Atmosphere

The absorption of solar EUV and x-ray radiation in the thermosphere provokes major changes in its composition as well as temperature: namely the ionization of atoms and molecules of air, which converts these neutral particles into electrically-charged ions (+) and free electrons (-).

The natural tendency of oppositely charged particles that come in contact with each other is to immediately recombine, and in the process, restore neutral atoms and molecules. But in the rarefied air of the upper atmosphere where encounters are less frequent, the newly-created electrons and ions can survive for a longer time: to about one second in the less diffuse air of the mesosphere to about an hour at the top of the thermosphere.

Although these added times are very short, they are long enough to establish and sustain an electrically charged layer, called the ionosphere, as a permanent feature of the upper atmosphere.

This region of electrically-charged particles extends upward from the middle of the mesosphere, about 35 miles above the surface, through the entire thermosphere and well beyond it, with a maximum density of charged particles in the region between about 125 and 375 miles altitude, which puts it within the lower portion of the magnetosphere. But it stretches on and upward, gradually decreasing in density with height, for thousands of miles.

While always there, the ionosphere is highly variable from place to place and time to time, in both structure and density. Dramatic changes occur each day when the Sun rises or sets. The ionosphere is also highly responsive to changes in the x-ray and ultraviolet emission from the Sun, which vary by more than a factor of two in the course of the 11-year solar activity cycle, and by even more on shorter time scales in the course of solar flares and the effect of solar rotation. It is also subject to magnetospheric storms, to the passage of meteors, and to the violent winds that sweep through the thermosphere on the heels of thermospheric temperature changes.

The ionosphere is made up of fairly distinct horizontal layers of differing density. When night falls and the Sun's electromagnetic radiation is turned off, the lower portions of the ionosphere—which are called the D and E *layers*, or *regions*—essentially disappear, leaving only the higher and denser F region—which survives, though much depleted, through the night, chiefly by the return of some of the electrons and ions from the magnetosphere.

During the day the F region is a major *exporter* of ions and electrons, which are carried upward into the realm of the magnetosphere by the thermal expansion of the ionosphere. During the night, as the ionosphere cools and contracts, some of these particles are drawn back to it, helping sustain the now depleted F layer until dawn: when, energized by the rising Sun, it bounds back to full strength again.

⊙ ⊙ ⊙

The existence of an electrically-conducting region in the upper atmosphere, though long suspected, was not confirmed until the age of radio, early in the last century when its reflective properties were employed for practical use in the long-distance transmission of radio waves. The reflectivity is greatest for lower frequency (longer wavelength) radio signals, for which the ionosphere serves as a mirror high in the sky to redirect upward-transmitted waves by a single forward reflection—or multiple reflections between the mirror and the ground—around the curvature of the spherical Earth.

Ionospheric reflection of radio waves was first put to use in 1901, by Guglielmo Marconi, with the successful transmission of a brief wireless signal from England to Newfoundland. But it was not until the 1920s, with the advent of broadcast radio, that the ionosphere was widely acknowledged and given its name. The fuller exploitation of all parts of the radio spectrum since that time has demonstrated that hour-to-hour and day-to-day variations and inhomogeneities in the ionosphere can affect all kinds of radio communications, from the lowest to the highest frequencies, including those used today in connection with the communication satellites that are more and more employed for television, satellite phones, GPS, and national security.

⊙ ⊙ ⊙

And so it was that a very old, unknown and invisible feature near the top of the Earth's atmosphere came to be of immense economic and practical importance on the ground.

The End of the Atmosphere

Unlike the oceans, the Earth's atmosphere holds claim to no fixed upper bound. It comes to an end instead through a process of gradual diminution: with the air growing ever thinner and more diffuse the higher one goes, until it eventually blends into the almost-vacuum of near-Earth space, much as recorded sound is faded into silence at the finish of a song.

The imaginary line that marks the threshold of *space*—like the invisible Equator ceremoniously crossed by cruise passengers—is for astronauts arbitrarily taken to be 60 miles above the surface of the Earth, where the pressure and density of air have already fallen by more than a factor of a million. This puts this dividing line in the lower thermosphere, toward the bottom of the ionospheric E layer where neutral and ionized particles coexist, and within the part of the upper atmosphere where incoming meteors paint their star-like trails. But it is considerably lower than where manned spacecraft fly and far below the last vestiges of neutral air.

It was into this lower region of the thermosphere—from about 100 to 200 miles above the surface—that manned spacecraft were first placed in orbit about the Earth: piloted in 1961 by Yuri Gagarin and in 1962 by John Glenn. To reduce the effects of atmospheric drag, instrumented spacecraft with longer life expectancies were later launched into higher orbits, into the more rarefied air of the upper thermosphere and the region beyond it, the exosphere.

The *neutrally-charged* atmosphere comes to an end at about 600 miles above the Earth's surface, where electrically-charged ions and electrons—created by the action of solar EUV and x-ray radiation on neutral atoms and molecules—have become the dominant species.

The change from a *neutral* gas to a *charged*, plasma atmosphere begins in the high stratosphere, some 40 miles above the surface of the Earth. From that level upward—through the three variable layers of concentrated ions and electrons that make up the ionosphere—electrically-charged constituents are more and more plentiful. Behind this persistent shift is the gradual thinning of air with altitude, which (1) allows more of the full blast of short-wave solar radiation to reach the neutral particles; and (2) extends the time allowed to newly-created ions and free electrons before they re-combine and neutralize each other.

Where the thermosphere ends, the outermost portion of the atmosphere, or exosphere, begins: the last thin vestiges of the atmosphere where all particles—neutral or charged—are on their way out of the atmosphere, and hence the name. The few neutral atoms and molecules that are there escape on the wings of their high thermal velocities, some of them into the magnetosphere.

Into the Magnetosphere

In this vast region of near-Earth space, it is no longer pressure and heat and chemistry, but *magnetic* forces that hold the reins of power and control: trapping and confining electrically-charged particles that have exited the exosphere, while at the same time shielding the planet from many charged particles that come at it from other directions.

The trapping and confining are accomplished within a grandiose and many-chambered structure made of magnetic lines of force that reach outward from the planet's internal magnetic field.

As our planet's first and outermost line of defense, the magnetosphere deflects away from the Earth most of the charged, energetic atomic particles that stream outward from the Sun, and some that come from other cosmic sources. Its outer boundary, called the magnetopause, is the unseen line, drawn in space, that for billions of years has separated the Earth and its defenses from the hostile world of near-Earth space.

Although we never see it, the mighty fortress of the magnetosphere is truly immense: large enough to dwarf the planet it encloses and protects, extending outward from the Earth toward the Sun for a distance of roughly 40,000 miles, or ten times the radius of the Earth. Moreover in the opposite, anti-solar direction it stretches outward into space for at least 800,000 miles: more than three times the distance to the Moon. Were it as large as its voluminous magnetosphere, our little Earth would replace Jupiter as the largest planet in the solar system.

Inside the inner magnetosphere we are yet within the strong grip of the Earth's gravity, and still in the company of some un-ionized atoms and even molecules of what might be called air, were they not so few and far between. As noted above, these last remnants of oxygen and nitrogen and other elements will continue to be found, though in ever-dwindling numbers, for hundreds of miles above the lower boundary of the magnetosphere.

In this fast-fading part of the gaseous atmosphere, neutral atoms of different elements are sorted by weight based on their response to the pull of the Earth's

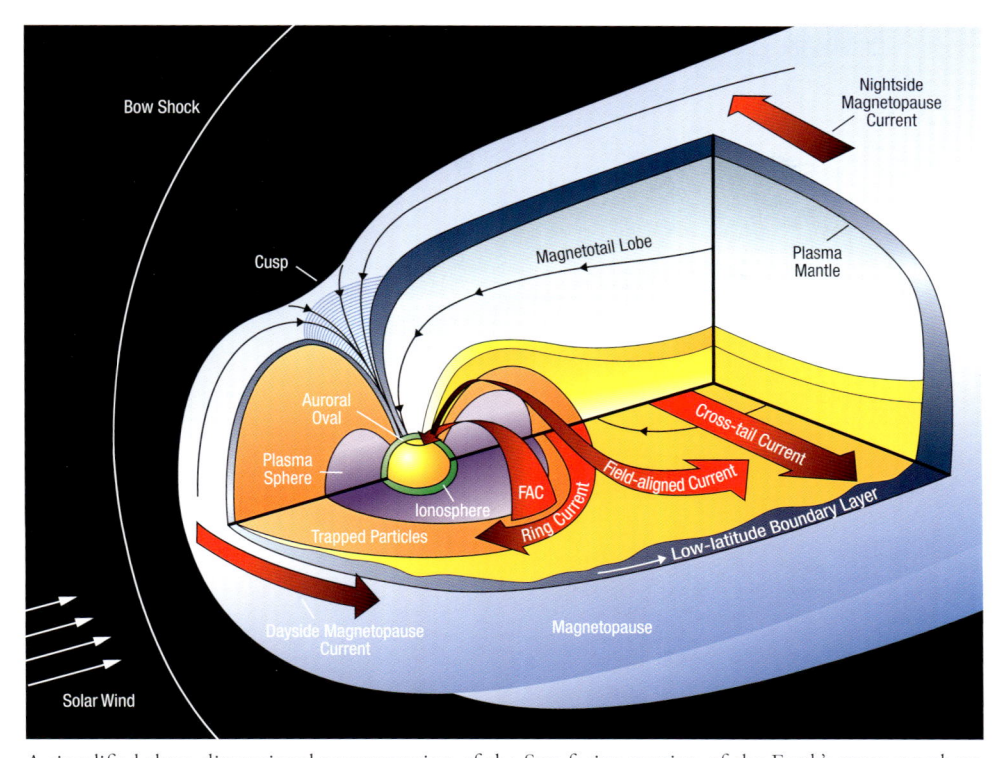

A simplified three-dimensional representation of the Sun-facing portion of the Earth's magnetosphere: the volume of space surrounding our planet that is dominated by the geomagnetic field and populated with plasmas from both the ionosphere and the solar wind. In the "upstream" direction, at left, the magnetopause, or outer boundary of the magnetosphere, is compressed by the solar wind on its Sun-facing side, pushing this boundary inward to about 40,000 miles (5 Earth diameters) above the planet's surface. Farther upstream a standing shock wave called the bow shock forms as the supersonic solar wind is slowed and heated by its initial encounter with the still-distant magnetosphere. On the night or "downstream" side, right, interaction with the streaming solar wind stretches the Earth's field into an elongated tail which extends more than 30 times farther: well beyond the orbit of the Moon and what is shown here. The ionized gases that populate the magnetosphere are extremely dilute and wholly invisible to our eyes. Nonetheless, they drive a system of powerful electric currents which during magnetic storms can dissipate well in excess of 100 billion watts of power, comparable to the output of all the electric power plants operating in the United States.

gravity. Through this process, heavier atoms and molecules are held closer and more tightly to the Earth, and the lighter more loosely bound. Thus as one moves higher in the atmosphere there are fewer and fewer heavy elements, until near the top, only hydrogen and helium, the lightest of all the elements, remain.

At that level many of the atoms that are present are no longer the gravitational captives of the Earth. Thanks to the energy and speed they acquired in their passage through the heated thermosphere, these fast-moving atoms are well on their way to freedom: no longer the wards of the little planet that has been their home for more than 4 billion years.

The Form and Function of the Magnetosphere

Long before Columbus, voyagers on land and sea had made use of the fact that the Earth was somehow magnetized: in such a way that one end of a lodestone or an iron compass needle was attracted toward the North Pole of the Earth, and the other to the South Pole.

As more was learned about magnets and the nature of magnetism, it came to be known that the Earth's magnetic field was a lot like that of a simple bar magnet: that is to say, with an oppositely-magnetized pole at either end, connected, one to the other, by curved invisible magnetic lines of force. Each of these was expected to arch outward from its roots in one of the magnetic poles of the planet and follow a gradual C-shaped curve, some distance above the surface of the Earth, to connect at the opposite magnetic pole.

It was also surmised, as early as 1930, that this idealized picture of the Earth's magnetic field—which should take a form similar to the pattern traced by iron filings above a laboratory magnet in a classroom demonstration—would be disturbed and distorted whenever charged particles shot out from the Sun impinged upon it.

Further insight into the likely form of the magnetosphere came in the late 1950s and early 1960s with the realization that a continually blowing solar wind would exert unrelenting pressure on the Sun-facing half of the magnetosphere. This was bound to hammer the idealized rounded C-shaped form of field lines on the day-side of the Earth into a flatter C, and in the process push the protective magnetic shield on that side much closer to the surface of the planet.

But it was not until the voyages of exploration of near-Earth space began, in the remaining years of the 20th century, that the actual form, function and great extent of the Earth's magnetosphere—and its ever-changing nature—came to be fully known and understood.

In many ways, including the societal value of what was found, these venturesome missions to explore and map the unknown near-environment of the Earth were not unlike the great voyages of discovery of the late 15th and early 16th centuries.

Most striking among their discoveries was the pronounced asymmetry of the magnetosphere: highly compressed and flattened on the side that faced the Sun, and grossly distended on the opposite, down-wind side. There, above the darkened half of the planet, the solar wind drags the lines of force of the Earth's magnetic field in the anti-solar direction into an extended tail, like that of a comet.

Within this grossly lop-sided magnetic shield, shaped by the Sun, the spherical Earth spins on its own axis at the rate of one rotation per day. The result is that the thickness of the magnetosphere above any fixed place on Earth is continually changing, and by vast amounts: thinnest at local noon; increasing through the afternoon until, as darkness falls, it stretches out above one's head beyond the orbit of the Moon, only to shrink back again to but a twentieth of its nighttime extent when at dawn the Sun again appears.

The Paths That Particles Follow

The Earth's magnetosphere, though an essential protector of life on Earth, is not 100% effective in this regard. Plasma and higher energy particles arriving from the Sun or other celestial sources can find their way, if their energies are sufficient or conditions are right, through the outer bulwarks of the magnetosphere and into the ionosphere and upper atmosphere.

As noted earlier, the incoming solar wind plasma, as it approaches the Earth's magnetosphere first comes upon a more distant, then a nearer barrier. The initial bulwark is a shock wave—known as the bow shock—which forms ahead of the Sun-facing "nose" of the magnetosphere.

There, pressure forces deflect, slow and heat the incoming plasma diverting it away from the direct Sun-to-Earth line. Similar forces produce the swept-back sheath of illuminated dust which forms around the solid head of a comet approaching the Sun.

Slightly farther on, the incoming plasma comes in contact with the magnetopause, the outer boundary of the magnetosphere. There it meets a barrier of closed magnetic field lines that are oriented more perpendicularly to the particle's path.

Most of the onrushing plasma is diverted around the magnetopause, like the diversion of a flowing stream by a partially-submerged rock. Some of the diverted flow sweeps over the poles of the Earth, and a fraction of that will find its way into the magnetosphere through unguarded gaps at the poles called the polar cusps.

Another means of entry for incoming plasma in either the solar wind or from more energetic particles is available in the vicinity of the equatorial, Sun-facing nose of the magnetopause. Whether that door is open or closed depends upon the polarity and orientation of the imbedded magnetic field that the alien plasma brings with it. When it is the same (or almost the same) as the orientation of the Earth's magnetic field, the particles that impinge upon it will be deflected and redirected as noted above, to continue their downward passage around the outer boundary of the magnetosphere.

If the polarity and orientation of the impinging magnetic field are opposite to that of the Earth (or nearly so) magnetic field lines in the incoming plasma stream will connect to the oppositely-directed field lines of the magnetosphere, which opens up additional routes, some through the polar cusps, into its guarded interior.

Captive Particles In the Magnetosphere

The same lines of force that prevent charged particles from entering the magnetosphere and atmosphere serve as well to entrap and confine any that from whatever source have found their way into the magnetic cage. These include some of the solar and cosmic particles that have worked their way around the bulwarks and into the magnetosphere, as well as those from the ionosphere and exosphere that have entered it from below.

Thus the electrons, protons, and ions that are magnetically confined at any time in the holding cells of the magnetosphere are often strangers: a mixture of immigrants arriving from afar and transients of local origin, who were caught on their way out of our atmosphere in the same magnetic net.

The most energetic are captured solar protons and electrons with energies in the range of a million or more electron volts, and they behave like swarms of angry bees, once free to roam but now confined. These and others less energetic are held in the Earth's magnetic grasp within flat current sheets in the inner magnetosphere that circle the planet in the plane of the Earth's magnetic equator. Others are held in the stretched magnetotail, in the radiation belts, or held within a well-worn set of tightly prescribed paths defined by the arching form of individual lines of magnetic force that are rooted in the north and south auroral ovals.

Constrained in this way, captive particles in the inner magnetosphere trace out a ritual dance of helical spins, bouncing reversals, and longitudinal drifts which take them back and forth from pole to pole, over and over again, while drifting around and around the Earth: and all of it frenetically, at break-neck speed.

⊙ ⊙ ⊙

In the first of the prescribed motions, a moving charged particle that comes into the vicinity of a line of magnetic force is drawn by its influence to gyrate (or circle) around it, as fast as a thousand turns per second, while following its full length from one pole to the other. As the tightly-spiraling particle approaches either end of a closed field line, at high magnetic latitudes, it meets an ever stronger and more crowded polar field, where adjacent field lines are packed more closely together.

There, at the end of the line, the second step is taken. Faced with stronger and more closely-packed field lines, the downward-moving gyrating particle is reflected and repelled, as from a magnetic mirror, and spiraled back from whence it came. When the now-reversed particle approaches the other end of the same closed field line—at the opposite magnetic pole—the same thing happens again. And on and on it goes, on a wild ride: bouncing back and forth from one pole of the Earth to the other, tightly spinning as it goes, completing each pole-to-pole trip in but a few minutes time.

The third obligatory motion is an induced, slower drift in longitude that results from the curvature of the magnetic field lines and the diminished strength of the field with distance above the surface of the planet. The effect is to nudge the gyrating particle a little bit in longitude—an electron in the eastward direction, protons or other positive ions westward—each time it bounces. Repeated nudging pushes it, bit by bit, around the Earth, such that the pole-to-pole motions of the particle sweep over the entire surface of the planet, all in about one hour.

The paths of the hordes of particles engaged in these and other patterns of motion delineate the form of the enclosures in which captured particles are detained. One follows the framework of closed field lines, reaching from pole to pole, that define the basic form of the Earth's magnetic field. The other is a ring of current flowing around the magnetic equator of the Earth, which defines two bagel-shaped reservoirs that store charged particles in two belts of charged particles.

The Earth's Radiation Belts

These near-Earth reservoirs of magnetically-trapped particles are the inner and outer Van Allen radiation belts, which were unknown and largely unexpected until early 1958, when they were chanced upon by the first U.S. satellite, *Explorer 1*, and named for their discoverer, the late James A. Van Allen.

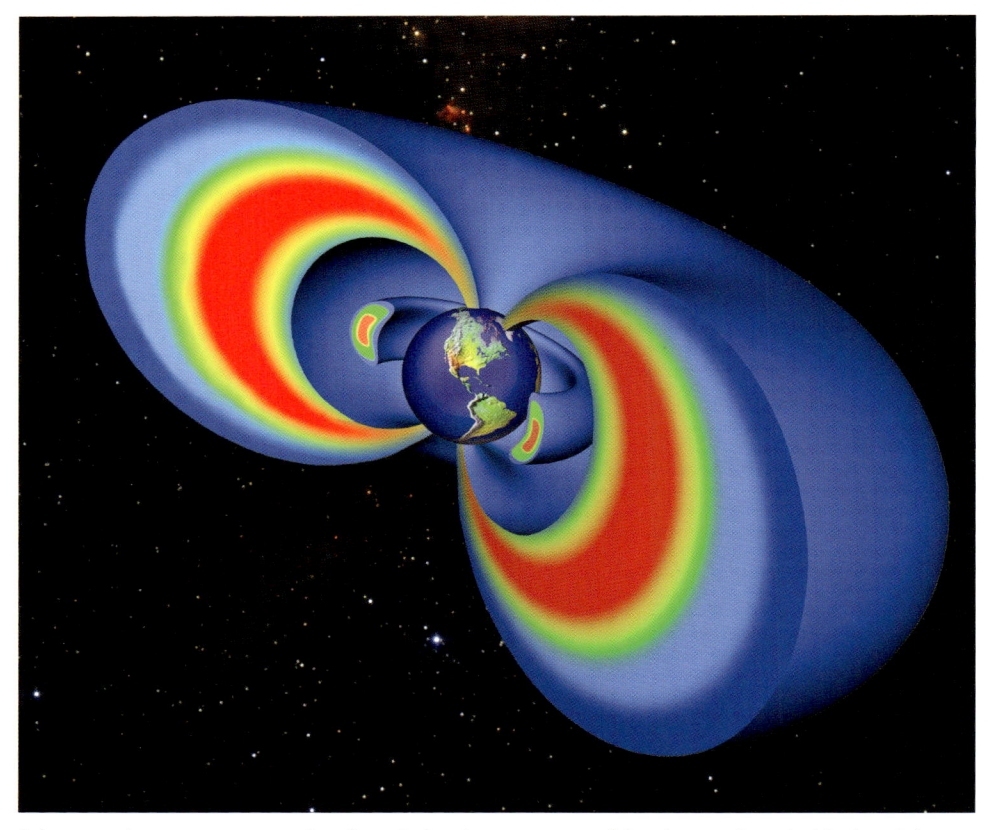

Schematic depiction, not to scale, of an idealized cross section of the plasmasphere (in blue) and the two doughnut-shaped concentrations of the high energy atomic particles that lie within its bounds, defining the Earth's inner and outer radiation Van Allen belts. Zones where the concentration of particles is greatest are colored red, then yellow and green. Between the inner and outer radiation belts is the (blue) gap or slot region, a zone swept free of particles by physical interactions.

Particles trapped within them are confined within two distinct and clearly separated zones, each shaped much like a bagel, that surround the Earth. The inner radiation belt—containing energetic electrons, protons and heavier ions, all in frantic motion—extends from the top of the atmosphere to a height of about 12,000 miles: a distance, measured from the *center* of the planet, of about three Earth radii, or 3 R_E.

The outer belt, of similar shape but larger diameter, reaches from about 16,000 to 24,000 miles (and at times as much as 36,000 miles) above the surface of the Earth, or between about 4 and 6 R_E, and at times as much as 10 R_E. Within it are the lightest and least energetic of charged atomic particles, which are weaker electrons with energies in the range of 10,000 to about one million electron volts. They come largely in lower energy plasmas that have entered the magnetosphere from the solar wind and the Earth's ionosphere.

Between the inner and outer belts—from about 12,000 to 16,000 miles above the surface of the Earth—is an empty gap or slot about 4000 miles wide: a region of near-Earth space in which protons but few if any electrons are found. This empty region between the highly-populated radiation belts is the result of a loss mechanism involving the interaction of charged particles and electromagnetic waves, that is particularly effective in that zone.

⊙ ⊙ ⊙

One of the ways by which moving charged particles can escape the radiation belts is through a chance encounter with any force that sufficiently alters its energy or disturbs its path and manner of movement. This can occur in the presence of electromagnetic radiation of a frequency that happens to resonate with the particle's own motion. This wave-particle interaction, which has maximum effect at a distance of 3 to 4 R_E, creates the gap between the inner and outer radiation belts: sweeping out newly-bound electrons as fast as they are supplied.

⊙ ⊙ ⊙

Few if any of the charged particles that were chanced upon in the Earth's radiation belts in 1958 are still there today, for although some can indeed remain for hundreds of years, following the same well-worn paths, most—in time—will find their way out of captivity, through the wave-particle process just described, or other mechanisms.

Among the other processes that continually deplete the particle belts are collisions between the charged particles that spiral downward with atoms and molecules of air at the foot points of magnetic field lines high in the atmosphere, other forms of downward diffusion, and almost any externally-caused disturbance of the magnetosphere. Particles to replace the departed continually arrive from three quite different sources: the Sun, the Earth's ionosphere, and some of the energetic particles called cosmic rays.

Cosmic rays that come from outside the solar system arrive with energies per particle of up to a billion or more electron volts: sufficient to speed right through the magnetosphere as though it weren't there. Those found in the inner radiation belt are their less energetic "daughter particles"—a second generation of ions created in the Earth's atmosphere when the original primary cosmic ray particles collide with atoms and molecules of air. Some of these secondary cosmic rays can move upward by diffusion through the upper atmosphere and into the magnetosphere, where, if they carry an electric charge, they are entrapped.

These upwardly-mobile secondary cosmic rays serve also as a primary source of particles for the inner radiation belt.

Charged particles are also created in the mesosphere and thermosphere when neutral atoms and molecules of air absorb far ultraviolet and x-ray radiation from the Sun. Upwardly-mobile ions and electrons from the upper thermosphere are subject to the same fate of magnetic capture and confinement as those produced by incoming solar particles and galactic cosmic rays.

⊙   ⊙   ⊙

Test explosions of nuclear weapons in near-Earth space—at a time when they were allowed—were also found to add energetic charged particles to the Earth's radiation belts. Best remembered is the *"Starfish Prime"* test conducted by the U.S. over Johnston Island in the Pacific Ocean on July 9, 1962: when an awareness of the radiation belts was but four years old.

THE EARTH'S RADIATION BELTS

FEATURE	GEOCENTRIC DISTANCE IN R_E	DISTANCE ABOVE EARTH'S SURFACE IN MILES
Inner Radiation Belt	1.2 - 3	650 – 12,000
Slot Region	3 - 4	12,000 – 16,000
Outer Radiation Belt	4 - 6+	16,000 – 24,000+
Plasmasphere	1.2 - 6	1000 – 20,000
For Reference: Orbits of GPS Satellites	5	≈16,000
Geosynchronous Orbit	6.6	22, 200

Exploded in the thermosphere at an altitude of 250 miles (about the same height at which the *International Space Station* now operates), the *Starfish* 1.4 megaton thermonuclear device filled the radiation belts with highly energetic electrons and positrons, some of which were still there five years later. The consequences of this apparently unanticipated happening were more than academic, for these entrapped particles of human origin crippled about one-third of all spacecraft in low-Earth orbit at the time, including *Telstar*, the first commercial communications satellite.

The Plasmasphere

Co-existing with the swarms of high energy particles held captive in the inner and outer radiation belts is another more extensive torus, depicted in blue on the figure on page 88, called the plasmasphere, which is filled with charged particles whose energies are many orders of magnitude weaker. Particles found in the plasmasphere are electrons, protons and singly ionized helium, along with a small amount of singly-ionized oxygen. These are for the most part remnants of the splitting apart of these atoms by solar short-wave radiation far below in the thermosphere and mesosphere. Atmospheric diffusion carries them along magnetic field lines into the magnetosphere, where they are captured and held.

The energy of any one of these more lethargic or "cold plasma" particles is at most a few electron volts, which for comparison corresponds to the thermal motion imparted by being heated to 20,000° to perhaps 40,000° F. In contrast, the electrons and ions in the hotter plasma of the inner radiation belt are suprathermal particles with energies of up to a million or more electron volts, corresponding to almost unimaginable temperatures of billions of degrees.

The plasmasphere extends from just above the ionosphere—about 1000 miles above the surface of the Earth—to a distance of up to 20,000 miles, or 6 R_E, which is almost a tenth as far as the Moon. As such, the plasmasphere overlaps the full extent of the inner radiation belt (about 1¼ to 3 R_E) and a good part of the outer one (4 to 6 R_E).

The cold plasma in the plasmasphere is held there by the same lines of magnetic force that confine hotter plasma in the radiation belts. But since these particles are less energetic, they are bound by a different set of constraints.

One difference is that they are spread continuously and without interruption from the exosphere to about 6 R_E, for they are unaffected by the "zone of avoidance" mechanism that sweeps energetic electrons from the gap in the radiation belts between 3 and 4 R_E.

As a result, the extent of the plasmasphere on any day, and how densely it is filled, are more variable than the radiation belts. Particle density can vary from about 1000 to 20,000 captive protons per cubic inch. The outer boundary of the plasmasphere, called the plasmapause, can reach all the way out to the magnetopause. And like the radiation belts, the plasmapause retreats inward toward the surface of the planet at those times when the Earth's magnetic field becomes connected to that of the Sun, moving at times from a distance of about 20,000 to but 4000 miles above the surface of the Earth. Like the magnetosphere of which it is a part, the plasmasphere is also asymmetric, bulging outward on the nighttime side.

Because of their low energies, particles in the plasmasphere pose no hazard to spaceflight, in contrast to the swarms of charged particles that encircle the planet in the radiation belts which can affect manned or instrumented spacecraft that enter or pass near them. Manned spacecraft that routinely travel in orbits nearer the surface of the Earth—like those that took the first cosmonauts and astronauts around the world in ninety minutes, or the well-worn trails of the *International Space Station*—are largely exempt, for these venture no farther than a few hundred miles above the surface of the Earth, well within the thermosphere and ionosphere.

The Heliosphere

When spacecraft cross the outer boundary of the magnetosphere they leave behind all natural protection against high-energy atomic particles from the Sun or the cosmos. From this point onward, and for at least the next 15 billion miles, spacecraft—manned or instrumented—will travel in a much harsher environment: the vast untamed realm of the much extended Sun, called the heliosphere.

This immense zone in interstellar space is the region dominated by the solar wind and the remnant parts of the Sun's magnetic field that are carried with it, and it stretches outward from the Sun to a distance well beyond the world of planets and planetesimals. Beyond its distant outer boundary—called the heliopause—lies the broader realm of other stars and other galaxies.

The heliosphere is commonly described as an oversize "bubble" surrounding the Sun and all the planets: a protected zone within the interstellar plasma, carved out by the solar wind and sustained by the Sun's extended magnetic field. It "protects" in the sense that the solar plasma that flows continually outward from the Sun is strong enough to fend off most of the plasma that comes in stellar winds from other stars, and to keep out all but the most energetic cosmic rays.

In this sense, the heliosphere is a much larger version of the Earth's magnetosphere, which provides a smaller bubble of magnetic protection around our tiny planet, protecting life upon it from most incoming charged atomic particles, both from the Sun and from more distant cosmic sources. The mighty Sun would seem to require no such protection for itself from incoming atomic particles—whatever their energy—although the planets, all of which are found within the innermost core of the heliosphere, benefit from the presence of this distant outer rampart and the additional magnetic shielding that it provides.

Within the heliosphere the solar wind blows radially away from the Sun and in all directions. It does so until the solar plasma has traveled so far that the outward pressure it exerts, determined by its speed and its ever decreasing density, no longer exceeds the competing, combined pressure of similar stellar winds from all the other stars, near or far.

⊙ ⊙ ⊙

The distance at which the solar wind first meets and is overcome by the pressure of incoming stellar winds varies with direction. It is nearest on the side that faces the direction in which the Sun and the heliosphere are moving through the interstellar medium: which indeed they are, taking the Earth and all the other planets with them. The velocity of the Sun relative to the local interstellar medium in our region of the Galaxy is a little more than 12 miles per second, or 45,000 miles per hour. And its direction—should you want to check it out on a starry night—is toward the bright star *Vega* in the summer constellation *Lyra*, just north of the Milky Way.

At the place where the solar wind first begins to feel the competing force of stellar winds a shock wave forms, much like the bow shock that forms near the outer boundary of the Earth's magnetosphere, or the bow wave ahead of the bow of a moving ship. In passing through this termination shock, the solar wind slows down from supersonic (about a million miles per hour) to subsonic speeds, and is partially deflected away from the direction of the Sun's motion through the interstellar medium.

The heliopause—marking the true outer limit of the heliosphere—lies beyond the termination shock and is separated from it by the sheath of weakened and diverted solar wind. At its closest point, directly ahead of the moving Sun, the distance between the termination shock and the heliopause is a billion miles or so. This places the end of the heliosphere at a minimum of perhaps 10 billion miles from the Sun, or roughly 30 times more distant than Pluto. Thus the Earth and all the other planets and planet-like objects occupy but the inner core of the far larger heliosphere.

The shape of the heliosphere, like that of the magnetosphere, most closely resembles an extended windsock, with a long tail that stretches behind the moving Sun as it travels through space. It is also thought that the extent of the heliosphere may respond to solar activity in that some *in situ* evidence suggests that the termination shock lies farther from the Sun in years when the Sun is more active and closer in years of minimum activity.

In this case the response at the termination shock to changing conditions on the Sun will be necessarily delayed by the year or so needed for the solar wind to reach that far-distant point.

Cruising the Heliosphere

More than 3,000 spacecraft—an average of more than 100 each year—have been successfully lifted into near-Earth space since the launch of *Sputnik* in the autumn of 1957. Only a handful of these, however, have ventured beyond the magnetopause: the line that marks the outer limit of our world and the beginning of the heliosphere. Most of the rest have been placed in more mundane, circular orbits around the Earth at varied angles of inclination to the plane of the Earth's equator, including many whose orbits take them in polar orbits over the North and South Poles of the planet.

Most of these Earth-bound spacecraft, including John Glenn's *Mercury 6* and today's *International Space Station*, have followed what are now familiar courses around the planet while clinging closely to it: remaining within the bounds of the upper thermosphere at altitudes of several hundred miles and completing each pass around the Earth in about ninety minutes time.

Some—including almost all communications satellites and the weather satellites that provide the familiar images of cloud cover for the evening news—are placed in higher, geosynchronous or geostationary orbits for which the time needed to circle the Earth is precisely twenty-four hours: the same as the rotation rate of the Earth itself. Satellites that orbit the Earth in this way appear fixed in the sky—unlike the ever-circling Sun and Moon and stars—remaining night and day above the same fixed geographical region on the surface of the Earth.

To achieve a geosynchronous orbit, a spacecraft must be lifted well above the thermosphere to an altitude of 22,200 miles: which is almost a tenth of the distance to the Moon. At this height, it will circle the Earth within the outer radiation belt, still within the magnetosphere but tens of thousands of miles below its outer boundary.

Among those that have gone beyond that line are the manned and unmanned spacecraft that have traveled to the Moon, instrumented explorers of the solar system and its planets, and "*in situ*" samplers and monitors of the solar wind. With very few exceptions, however, all of these ventures into the open space that lies beyond the Earth's protective shell have been limited to the two dimensions of the ecliptic plane in which the Earth orbits the Sun. This restriction has kept these venturers within a very thin slice of the true extent of the three-dimensional heliosphere.

⊙ ⊙ ⊙

The most venturesome of those that have left the ecliptic plane is the *Ulysses* spacecraft, launched in the autumn of 1990, and first dispatched to distant Jupiter. There, almost two years later and far from home, it succeeded in harnessing the gravitational pull of that most massive of the planets to hurl itself, as from a slingshot, out of the ecliptic and into a polar orbit around the Sun, passing alternately over the north and south rotational poles of the star: where no spacecraft from the Earth had ever gone before. And there it operates today.

This truly epic voyage of discovery takes *Ulysses* about as close to the Sun as the orbit of Mars, and as far away as the orbit of Jupiter: in an elliptical orbit that takes more than six years to complete. Along the way, month after month and year upon year, it gathers and sends back to Earth a host of *in situ* measurements that tell of conditions in the solar wind, probing the three-dimensional inner heliosphere through all seasons of the Sun's 11-year cycle of activity. These data have proved to be of inestimable value in defining the velocities and the solar sources of both high- and low-speed streams in the solar wind, the three-dimensional nature of the Sun's extended magnetic field, and the spatial distribution of cosmic rays in the inner heliosphere.

⊙ ⊙ ⊙

Two other pioneering spacecraft, *Voyager 1* and *Voyager 2*, launched about two weeks apart in 1977—thirty-two years ago—have now made their way to the outer limits of the heliosphere. Within perhaps the next ten years the most distant of the pair, *Voyager 1*, is expected to achieve the long-sought goal—like that of finding the source of the Nile, or first reaching the frigid Poles of the Earth—of finding the end of the heliosphere: the heliopause. And crossing it, to enter the interstellar medium.

These identical spacecraft were sent like two ships of discovery on extended voyages through the solar system. On the way *Voyager 1* first visited the giant planet Jupiter, and *Voyager 2*, not far behind, gaseous Saturn, Uranus and Neptune, as well as their moons, in part to take the measure of their magnetic fields. By 1989 they had made their last planetary encounter; accomplished all their initial tasks and more; and sailed on to explore the farthest limits of the solar system. Because the radiation from the Sun is so feeble at these great distances—where it appears as but a point of light in a black sky—the Voyagers were both powered by batteries continually charged by the decay of plutonium, which are expected to last until about the year 2020.

On the 16th of December of 2004, after twenty-seven years of voyaging, *Voyager 1*, then about 8.7 billion miles from the Sun and the Earth, came upon and passed through the termination shock of the heliosphere: where the solar wind first experiences effects of opposing gusts from other stars. Like the reflected swells that told early Polynesian navigators that land lay ahead, somewhere over the horizon, this expected wall of high pressure in the farthermost heliosphere served as a clear sign that the heliopause—the true end of the heliosphere—lay out there, somewhere, dead ahead.

The news of this historic crossing, coded in solar wind and magnetic data that were transmitted—like all radio waves—at the speed of light—took thirteen hours to reach the Earth. In the first year of its journey, as *Voyager 1* passed the Moon, the transmission time was less than two seconds, and from Jupiter, forty minutes.

Voyager 1 now travels onward within the [heliosheath](#) of diminishing solar winds that stands between the termination shock and the end of the heliosphere, with miles to go before it sleeps. The distance it must still travel to make it there is not known with certainty, but thought to be another billion miles or so. At its present rate of travel it will require almost another decade—until about 2015—to reach that goal.

Voyager 2, following the same course at the same speed is about two years behind it, soon to make its own passage through the termination shock, and to cross the finish line in about 2017. About three years later—after sailing another half billion miles into the Great Unknown—the atomic batteries which by then will have kept the little spacecraft alive for more than forty years will finally fail, leaving the then voiceless ships to sail on alone in silence, into the dark of space.

Radiation in the form of heat and light provides almost all of the energy transferred between the Sun and the Earth, shown here in close proximity. The diameter of the Sun is actually about 100 times larger than that of the Earth, and the distance that separates them—about 93 million miles—is equal to well more than 100 solar diameters. Thus the Earth seen from its constant benefactor would appear little larger than a dot in the sky.

FLUCTUATIONS IN SOLAR RADIATION AT THE EARTH

Changes in Total Solar Irradiance

It has long been known that the face of the Sun changes from day to day with the coming and going of sunspots and faculae on its white-light surface; that the outer layers of the solar atmosphere—the chromosphere and corona—are even more dynamic and variable; and that solar explosions and eruptions are everyday occurrences on the star. But to what extent do these transient features alter the amount of solar energy we receive at the Earth, 93 million miles away?

For example, how great of a surge can we expect when the flood of x-rays and extreme ultraviolet radiation from an explosive solar flare arrives at the Earth, in eight minutes time? Or seven minutes after that, when solar energetic protons, propelled outward at half the speed of light, come upon us from the same cataclysmic event on the Sun? How much added energy must the outer atmosphere of the Earth absorb when an incoming blob of charged solar plasma, ejected from the corona a few days before, becomes magnetically connected—like the accidental contact of two hot wires—to our own magnetosphere?

Or the most commonly-asked question regarding the Sun (which remained unanswered and unanswerable for more than 360 years after Galileo first found sunspots): What is the effect— if any—on the total amount of heat and light we receive when sunspots come and go?

☉ ☉ ☉

Since sunspots appear darker than the surrounding photosphere, they have obviously reduced the flow of radiation from the restricted area of the solar disk that each one covers. Faculae, which are brighter than the surrounding surface, clearly add locally to the total radiation emitted.

Could it be that what one takes away, the other immediately restores? That each sunspot simply acts as a plug to deflect without stopping the flow of solar energy that pushes inexorably upward from deep within the star?

Were this the case, the total solar irradiance—the amount of energy that the Sun delivers in all wavelengths at the top of the Earth's atmosphere—might indeed be called, as long it was, the "solar constant."

We know now, based on more than a quarter century of direct measurements taken from space, that there is and never was a solar constant. The total amount of energy emitted from the Sun varies on all scales of time, from seconds to years, as does the energy emitted in any of the different spectral components that contribute to that sum. Everything changes, all of the time: the radiative energy emitted in x-ray, ultraviolet, visible, infrared, and radio wavelengths as well as the energy carried outward from the star in the form of energetic atomic particles. All of them respond to varying magnetic activity on the surface of the Sun.

We also know that the flow of solar energy in each of these components—and their sum—increases when the Sun is more active, peaking at times of maxima of the 11-year solar cycle, only to fall back down again in years of minimum activity.

☉ ☉ ☉

The overall level of *solar activity* varies as well on decadal periods of time, changing by a factor of five or more from one 11-year cycle to the next, when measured in terms of the numbers of sunspots seen on the Sun in the years around the maximum of each solar cycle. There are weak cycles and strong cycles, and prolonged periods in which the cycles are (1) persistently stronger (as during much of the last century); (2) weaker (as in the first three decades of the 19th century and at the turn of the 20th); and (3) almost but not quite absent altogether (as during the 70-year period that ended in 1715.) There is also some evidence of a longer solar activity cycle of about 80 years and another of more than a thousand years.

What is still not known and needs to be found is the extent to which these longer-term changes in solar behavior—spanning periods counted in decades to centuries—affect the total energy that the Sun emits, since the period for which we have actual measurements is as yet so short.

Precision radiometric measurements of the total heat and light and other radiation received from the Sun can be made only from above the Earth's atmosphere. This technically-challenging task was initiated on a continuing basis in 1978 and has been sustained since that time by a series of dedicated instruments on different solar spacecraft. When properly intercalibrated and pieced together, these invaluable data provide a continuous day-to-day record of the Sun's most vital sign which now extends for more than a quarter-century.

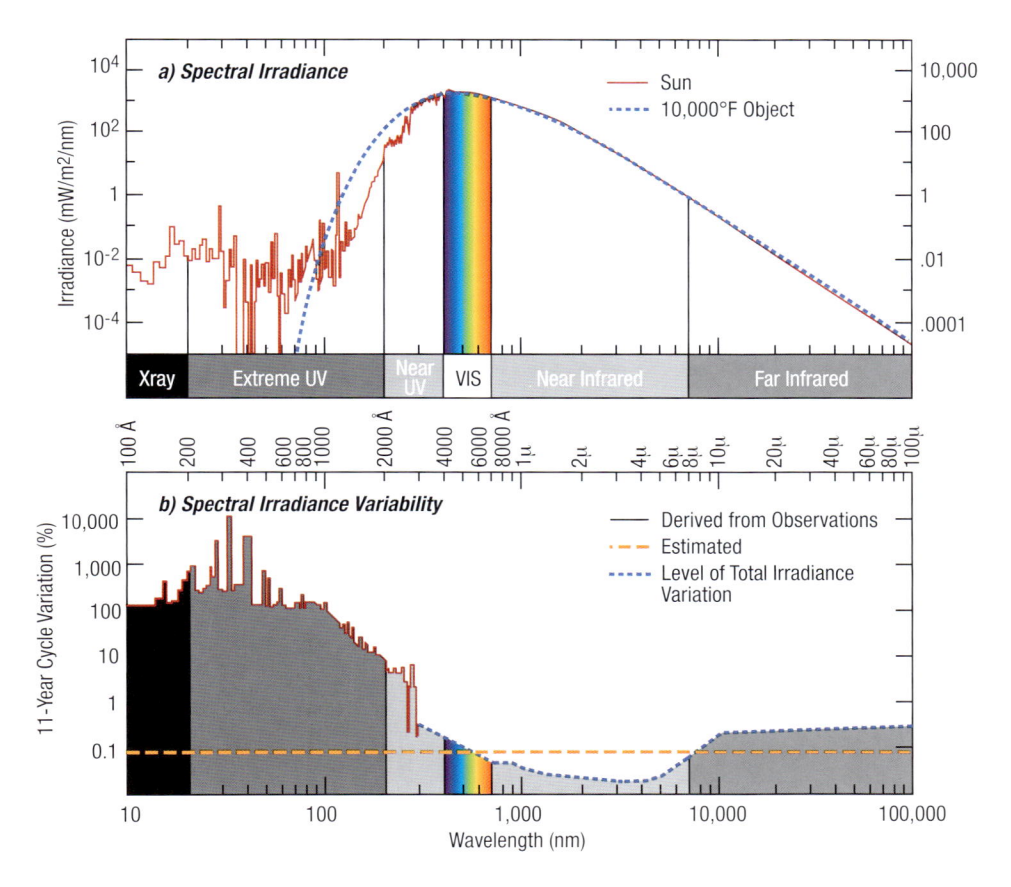

The spectral distribution of solar radiation and its variablility. (a) Upper figure: Red line: the amount of solar electromagnetic radiation in different wavelengths that falls on the top of the Earth's atmosphere, expressed in milliwatts per square per nanometer wavelength, compared with the relative amount of radiation (dashed blue line) expected from an ideal radiator at a temperature equal to that of the Sun's visible, photospheric surface. The excess solar radiation seen at the shortest wavelengths originates in higher regions of the solar atmosphere (chromosphere and corona) where the temperature is much higher. (b) Lower figure: the percentage variability of solar radiation in these wavelengths. In the visible and near-infrared spectrum the variation is at most about 0.1%, and only slightly more in the far infrared. The greatest variation is not surprisingly found in the short-wave radiation that emanates from the Sun's more volatile upper atmosphere. There the variation is several percent in the near-ultraviolet, 100% in most of the extreme ultraviolet and x-ray region, and up to more than 1000% in narrow wavelength regions corresponding to the wavelength-specific radiation (called spectral emission lines) of highly-ionized atoms, such as the very strong emission line at a wavelength of 30.4 nanometers (or 304 Ångstroms) coming from ionized helium.

Variability in Different Parts of the Spectrum

More than 90% of all the energy that leaves the Sun comes to us in the visible and near-infrared regions of the spectrum, in the form of light and solar heat. Both of these two components vary in step and in phase with changing solar activity: changing by at most a few tenths of a percent in day-to-day excursions and more slowly and systematically in the course of the 11-year solar cycle. Though small in terms of other climate modulators—such as cloud cover—that more forcibly and more randomly alter the amount of radiation received at the Earth's surface, these more persistent changes in external solar forcing are sufficient to leave discernible, 11-year marks on the surface temperature of both the lower atmosphere and the oceans.

Slightly less than 10% of the Sun's total output of energy comes to us in the form of near-ultraviolet radiation, which dictates the production of ozone and through that process heats the stratosphere. Since it originates at a higher and more active level in the Sun's atmosphere, solar radiation in this invisible part of the spectrum is more variable than the visible and near-infrared parts. Its response to changes in magnetic activity on the Sun is also much greater, varying by as much as 10%.

In addition, the photons that carry radiation in the near-ultraviolet are considerably more energetic and as such are potentially hazardous to life: with the potential of immediate damage to our eyes, skin and other human tissue, and more subtly to our immune system. Although most of the solar near-ultraviolet radiation incident upon the Earth is absorbed by ozone, a small fraction—more at higher altitudes and at lower latitudes—slips by the ozone filter in the sky and makes its way to the surface of the Earth to do its damage to those exposed.

The extreme ultraviolet (EUV) and x-ray radiation from the Sun arises from the progressively higher temperatures of the upper chromosphere, transition zone and corona. The structure and make-up of these upper reaches of the Sun are highly variable, as is the EUV and x-ray radiation from the Sun as a whole, fluctuating by an order of magnitude or more.

Solar EUV and x-ray radiation are also far more energetic than the near-ultraviolet and potentially more lethal. And even though these most energetic rays make up but one ten-thousandth of the total energy emitted by the Sun, they are sufficient to exert almost complete control of the thermosphere and the ionosphere.

SOLAR ENERGY RECEIVED AT THE EARTH

FORM	TOTAL ENERGY IN WATTS PER ACRE AT THE TOP OF THE ATMOSPHERE	FRACTION THAT REACHES THE SURFACE	FRACTION OF TOTAL ENERGY RECEIVED	VARIABILITY
TOTAL IRRADIANCE	5500	60%	~100%	0.1%
Near infrared	2800	55%	51%	0.05%
Visible	2200	75%	40%	0.1%
Near-ultraviolet	490	40%	9%	1%
Far-infrared	10	20%	0.14%	0.1%
Far-ultraviolet	.5	0	0.01%	15%
EUV and x-ray	.2	0	0.005%	up to 200%
PARTICLES				
Solar protons	0.002		negligible	up to x100
Solar wind particles	0.0003		negligible	up to x30
Galactic cosmic rays	0.000006		negligible	20% due to solar modulation

The Sun sends out radio waves as well, and in all directions, which can be clearly heard using a directed antenna as crackling static across the full span of the radio frequency spectrum. But the radio emission from the Sun is of little consequence in terms of direct terrestrial impacts since it carries so little energy, even during explosive flares, when the Sun's radio emission increases by many orders of magnitude. What the Sun emits in these very long wavelengths is, nevertheless, of considerable value as a diagnostic tool for remotely probing and monitoring the solar corona and transition zone, where radio frequency radiation originates, and as an alternate index of solar activity.

The largest spikes in the near-ultraviolet, EUV and x-ray radiation from the Sun come from the eruption of explosive solar flares, in which concentrated magnetic energy is suddenly and catastrophically converted into heat and light and short-wave radiation.

At these times the flaring region on the solar disk brightens dramatically in all parts of the spectrum, with the greatest increase in the shortest wavelengths. In EUV and x-ray radiation the burst of radiation is so intense that it dramatically increases the total radiation in these wavelengths received from the entire disk. In white, visible light this is almost never the case, and when it does—during the very rare occasions when a so-called white-light flare occurs—the brief increase in the total visible radiation from the Sun is no more than 1%.

Although the energy released at the site of the flare can equal the detonation of a tightly-clustered pack of 100,000 atomic bombs, the direct effect on the total amount of solar energy received at the Earth—a very small target at a very great distance—is to add, but briefly, about four watts to the more than 5540 kilowatts per acre that the Sun continually deposits in the form of light and heat.

Effects of the Sun's Rotation

The 27-day rotation of the solar surface has no effect on the total amount of radiation the Sun sends outward in all directions. But it plays a major role in modulating the amount the Earth receives by exposing us each day to a different part of the variegated surface of the radiating photosphere.

Were the Sun to spin much more slowly and, like the Moon, keep the same half of its surface turned always toward us, the energy we receive would still vary in response to both eruptive events and the birth and evolution of sunspots and faculae, which come and go in matters of days to weeks.

Solar rotation superposes on these intrinsic solar changes a more rapid and complex modulation, as new and different patchworks of sunspots and faculae are rotated into and then out of our view. In the process the Sun rolls out for us a panoramic display of its entire surface, continually updated and repeated for us—should we have missed all or part of the first show—fourteen times each year.

Were we unable to see the surface of the Sun we could still determine its average rate of rotation from ongoing observations of solar radiation received at the Earth in the energetic ultraviolet. As long-lived active regions are repeatedly carried across the face of the Sun by solar rotation these short-wave hotspots on the visible disk would appear in a recording of ultraviolet flux from the Sun like a sweeping beam of light from a lighthouse.

The effect on the ultraviolet radiation we receive introduces a related 27-day modulation in the amount of ozone in the stratosphere, which is largely controlled by short-wave solar radiation.

☉ ☉ ☉

The marks of solar rotation are also seen in continuous measurements of the total solar irradiance during the passage of large sunspots across the solar disk.

A large sunspot coming into view at the eastern or left-hand limb of the Sun will first appear very much flattened in one dimension due to the spherical shape of the Sun. If the spot is *circular*, it will first appear to us in a telescope as a dark vertical *line*, and in succeeding days, as more and more of its true area is turned toward us, as a flattened *oval*. As it moves with the Sun's rotation across the face of the Sun we are shown more of its true shape and size, which are fully revealed only when the sunspot reaches the central portion of the solar disk.

This imposed "foreshortening" effect can be quite evident in the record of total solar irradiance during the time that a large sunspot is carried by rotation across the face of the Sun. The amount of energy a sunspot takes from the total solar energy received at the Earth is proportional to its *projected area* on the solar disk: greatest when the spot reaches the central region of the disk and least when it lies near either limb. Thus as it is carried across the Sun by solar rotation a single large sunspot or sunspot group will often appear as a distinctive V-shaped notch in a continuous record of total solar irradiance.

Effects of the Earth's Orbit

A number of other influences, unrelated to the Sun itself, alter both the total amount of solar radiation the Earth receives, and how it is distributed over the spherical surface of the planet. Some of these, arising from the non-circularity of the Earth's orbit and a congenital tilt in its axis of rotation, continually modulate the amount of solar energy that arrives on any day, and any place.

Others, such as atmospheric absorption and reflection from clouds and aerosols—the middlemen in the radiative transfer business—take their often heavy cut from the amount that enters the top of the atmosphere as it streams downward toward the surface of the Earth. In the end, only about half of the all the energy that the Sun delivers on our doorstep—at the very top of the atmosphere—will ever make it to the ground.

☉ ☉ ☉

The fact that the rotational axis about which the Earth spins is not exactly perpendicular to the plane in which our planet circles the Sun introduces a dramatic annual modulation in the solar energy we receive. This considerable imperfection tilts the Earth's North and South Poles for half of the year *toward* the Sun and for the other half *away* from it. This produces a regular change in the height to which the Sun climbs each day in the sky, and through this a smooth annual variation in the number of hours of sunshine. From this tilt in the Earth's rotational axis come the seasonal rhythms which for eons have so strongly directed the course of life on the planet.

The magnitude of these seasonal changes obviously depends on one's latitude and the angle—currently 23½°—at which the Earth's axis is tilted. But unlike the globes on our desks and tables, the angle of inclination of the real Earth was not permanently fixed at The Factory.

The tilt of the actual, spinning Earth is nudged toward larger or smaller angles of inclination by the combined gravitational pulls of the other planets and the Moon. Because these change with time, the tilt angle slowly oscillates back and forth between limits of 22 and 25°, completing a full cycle in about 41,000 years. This too, has important climatic impacts since it affects how incident solar energy is apportioned over the spherical Earth.

⊙ ⊙ ⊙

Similarly, although we often think of the Earth's orbit as circular, the path we actually follow in our annual trip around the Sun is in truth slightly elliptical: an oval, not a circle, which in the present era takes us about 3 million miles nearer to the Sun at closest approach than half a year later, when we are farthest away. This annual variation of about 3% in the distance that solar radiation must travel to get to us—which ranges from 91½ to 94½ million miles—produces an annual modulation of almost 7% in the total radiation the Earth receives: an annual change that is a hundred times larger than any variation of purely solar origin.

The elliptical shape of our orbit is another consequence of the gravitational pulls of the Moon and the other planets on our own movement around the Sun. At present, the eccentricity or non-circularity the Earth's orbit—a measure of its departure from a perfect circle—is slightly less than 2%. But this, too, slowly varies with time, oscillating in the course of about 100 thousand years between nearly zero (when for a limited time our orbit is indeed a divinely perfectly circle) to an upper limit of about 6%.

⊙ ⊙ ⊙

The climatological impact of variable eccentricity is much affected by a *third* significant variable of our orbit about the Sun: namely, the *times* of the year when the elliptical path takes us closest to our parent star, and farthest away. These times of maximum and minimum intensity can fall in any part of the calendar, and do, following a cycle of about 22 thousand years. Like the other perturbations, it too is dictated by the changing pulls of the other planets and the Moon as these other bodies trace out their own celestially imperfect orbits.

In the present era the Sun is closest to the Earth in January and farthest in July. Thus the effect on the Northern hemisphere is to dilute, somewhat, the intensity of the seasonal changes that follow from the tilt of the Earth's axis. In the Southern hemisphere, approaching closer to the Sun in January (there, in early summer) and farther from it in July (in winter) has quite the opposite effect.

In the Southern Hemisphere the impact on seasonal climate of turning up the heat in summer and turning it down in winter is at present not as severe as in the Northern Hemisphere. There the effect is appreciably dampened by the current placement of the ever-drifting continents: specifically, the happenstance in the present era of more extensive oceans in the Southern hemisphere and the presence at the South Pole of a very large ice-covered continent, both of which provide thermal inertia.

☉ ☉ ☉

These slow and subtle changes in the orbit and inclination of our planet—known as the Milankovitch effect—act together to reapportion in time and reallocate in space the continuous flow of energy that the Sun delivers to the Earth. And though small in absolute terms, the internally-amplified climatic impacts of these changes appear to be sufficient to have served as a pacemaker for the coming and going of the major Ice Ages throughout the last million years of Earth history, as revealed in repeated patterns of ocean temperature obtained by paleo-oceanographers from the analysis of deep-sea cores.

Lost in Transit: The Fate of Solar Radiation in the Earth's Atmosphere

All else that modulates the amount of solar radiation that streams down upon the Earth—including all variability at the source itself and all short and long-term consequences of our orbit about the Sun—is dwarfed by what happens in the atmosphere itself: in but the last few hundred miles of a long, long journey. In its passage from the top to the bottom of the ocean of air, only a little more than half of the solar energy that falls upon the upper atmosphere will reach the solid surface and oceans of the Earth. The rest will be consumed as it speeds downward through the air, or reflected back towards the Sun.

About 30% of all solar radiation that arrives at the top of the atmosphere is immediately rejected by the planet and returned, unused, back into the dark of space: reflected mostly in the lower atmosphere by clouds and other suspended particles or at the surface by highly reflective water, ice and snow and different types of land cover.

The remaining 70% is put to use: the sum of the 50% that is absorbed by the surface to directly heat the planet and 20% that is selectively absorbed by atoms and molecules of air, mostly high in the atmosphere, or selectively scattered in the lower atmosphere to illuminate and color the sky. Included in the 70% is all the energy in the x-ray and extreme ultraviolet region of the spectrum, which is absorbed in the thermosphere and mesosphere; almost all of the near-ultraviolet, absorbed by ozone in the stratosphere and troposphere; and some of the infrared, absorbed in the same region by water vapor, carbon dioxide and the other molecular greenhouse gases.

All of these deductions vary from place to place and on all scales of time, as they together reduce and modulate the solar energy that ultimately reaches us at the bottom of the ocean of air.

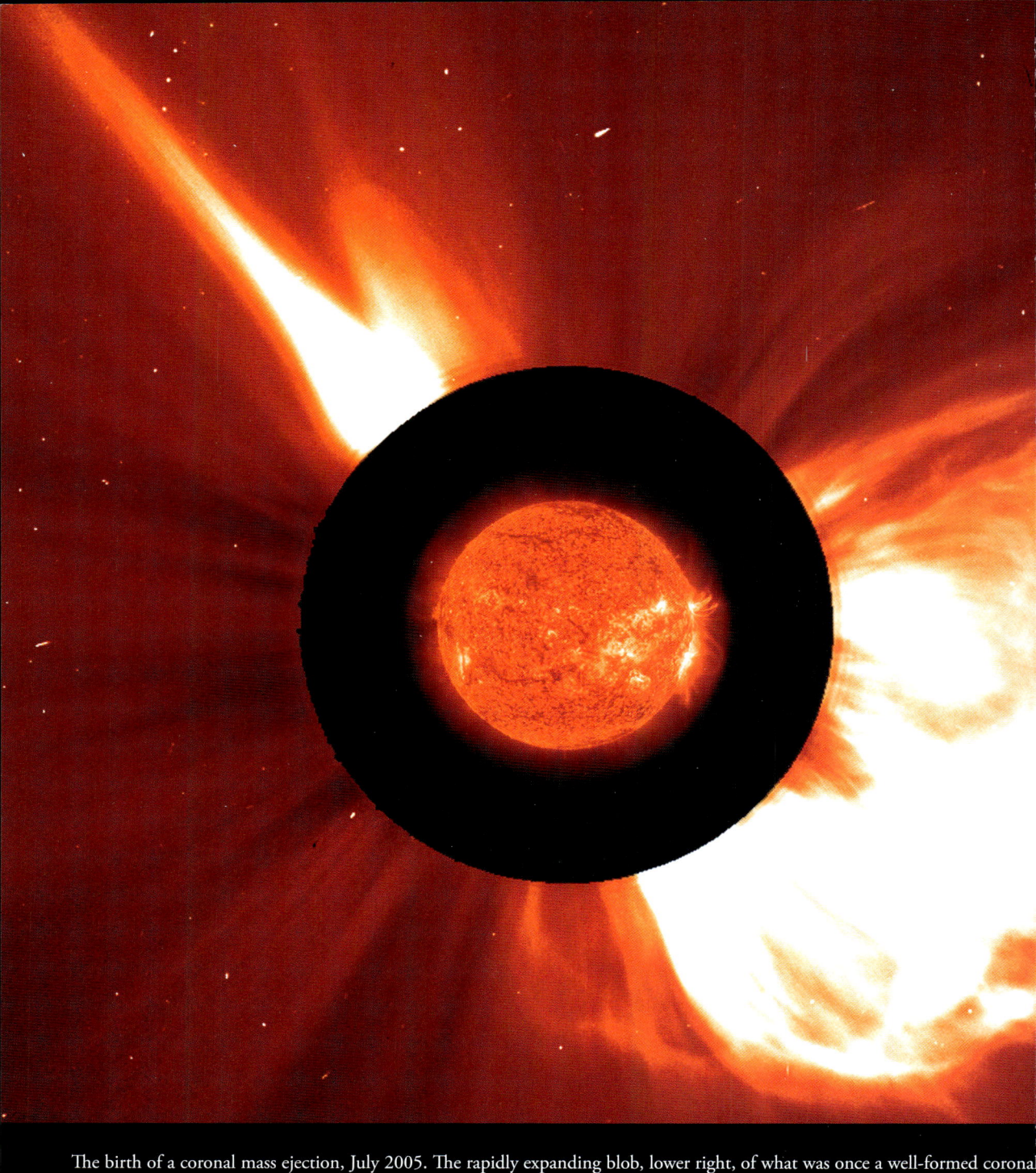

The birth of a coronal mass ejection, July 2005. The rapidly expanding blob, lower right, of what was once a well-formed coronal streamer (like that at upper left) is propelled outward, faster than a speeding bullet, into the heliosphere. This image of the Sun's outer corona was made in white-light by a continuously-operating coronagraph on the SOHO spacecraft. The inner image of the chromosphere, which identifies the relevant magnetically-active regions on the underlying surface of the Sun, was made at the same time and on the same spacecraft in the far ultraviolet. It has here been overlaid, to scale and properly oriented, on the black occulting disk of the corona, which blocks the solar disk and inner corona from our view. Violent CMEs of this kind are quite separate from the far gentler outward flow of less energetic particles that pour continuously outward from all parts of the Sun in the solar wind.

VARIATION IN THE FLOW OF PARTICLES AT THE EARTH

The Nature of Arriving Particles

Each atomic particle that leaves the Sun—in the solar wind, in CMEs, or from solar flares—is a small piece of the star itself, taken chiefly from the corona and transition zone. Because of the extremely high temperatures in these source regions, the atoms that escape the Sun have lost most or all of their electrons, yielding positively-charged ions and negatively charged electrons. For hydrogen—which is far and away the most abundant element on the Sun—the products of ionization are simply protons and freed electrons. For helium, the next abundant, and for all the other elements the products are free electrons and ions of different positive charges.

These newly-released particles, since they carry an electric charge, are at once bound to the strong magnetic lines of force that thread the solar atmosphere. Particles that leave the Sun in the solar wind carry these embedded magnetic fields with them: like missionaries sent on their way with bibles in hand. An assemblage of charged particles of this sort describe a state of matter—called plasma (a gas composed of ionized or charged particles)—whose behavior is quite different from that of the more familiar gaseous, solid, and liquid states of matter. One of these differences is the capability of retaining an imbedded or "frozen in" magnetic field.

The combination of charge and embedded field in solar plasma very much affects what happens when the ejected plasma comes in contact with the magnetic fields of the Earth or other planets.

Depending upon their energies, solar particles that arrive at the Earth produce quite different impacts. The least energetic are the solar wind plasmas, in either slow- or high-speed streams, as well as most particles carried Earthward in CMEs, which in either case consist of weaker, thermal electrons and ions. Their energies per particle (expressed in electron-volts, or eV) tell of the temperature in their region of origin in the chromosphere, transition zone, and corona of the Sun, and range from a few eV for electrons to at most at few keV ("kilo" or 1000 electron volts) for protons and other ions.

Highly accelerated (or suprathermal) particles include protons and other ions expelled more impulsively from flares and more commonly, particles in the solar wind plasma that have been accelerated en route to the Earth by fast-moving CMEs. The energies of flare particles are commonly in the range of MeV (millions of electron volts).

It is the considerable energy of these accelerated, suprathermal particles that produces significant terrestrial impacts and effects. These effects include, among others, splitting atoms and molecules apart in our upper atmosphere; creating and sustaining the Earth's ionosphere; and imposing direct hazards to human passengers in either air or spacecraft.

Galactic cosmic ray particles are more energetic, with typical particle energies in the GeV ("giga" or 10^9 electron volts) range, although their range of energies spans twelve orders of magnitude. Some of these incoming alien particles are endowed with energies so awesome—up to 10^{20}, or a hundred billion billion, electron volts per particle—that for most practical purposes they are unstoppable.

The speed of solar particles—a function of their mass and kinetic energy and in the case of the solar wind, the bulk motion of the plasma—also covers a wide range. A solar wind electron with an energy of 10 eV travels at a characteristic speed of about 1000 miles per second, to arrive at the Earth in a little less than a week. An energetic proton from a solar flare with an energy of 10^9 eV will travel a thousand times faster—at half the speed of light—covering the same vast distance in but 15 minutes. And because the proton is so much heavier, its potential impact on the Earth is immensely greater.

A stream of solar protons with energies in the range of 10^9 eV and moving at this near-relativistic speed constitutes a potentially fatal threat to anyone outside the atmosphere who should happen to cross its path, and as noted earlier, an almost unavoidable one, for they can penetrate four inches of lead.

ENERGIES AND CORRESPONDING TEMPERATURES OF SOLAR ATOMIC PARTICLES

PARTICLE ENERGY IN ELECTRON VOLTS	CORRESPONDING TEMPERATURE (°F)	TYPICAL SOURCE REGION
0.5	10,000	Solar photosphere
1	20,000	Low chromosphere
10	200,000	Transition zone
100 eV to 1 keV	10^6 to 10^7	Corona
MeV to GeV	10^{10} to 10^{13}	Solar flares

Solar Sources

Energetic atomic particles from the Sun begin their outward journeys from a variety of starting points in the solar atmosphere: some from eruptive prominences and active regions; some from coronal holes; and others from long coronal streamers. Some leave from low latitude regions on the Sun and others from its poles. But in every case their point of origin and direction of flow are defined by local characteristics in the Sun's magnetic field.

The most energetic streams of particles that reach the Earth arrive from either explosive solar flares or more often, from shock waves created by fast CMEs in the solar wind plasma. The energy that initiates and propels particles outward from flares comes from the sudden conversion of some of the magnetic energy in solar active regions into thermal and kinetic form. These short-term, impulsive blasts of highly energetic particles are superposed on top of slower and less energetic solar wind plasma.

What we sense at the Earth is a slowly-changing background of galactic cosmic rays on which fast and slow streams of solar wind plasma are superposed. On top of these are more sporadic episodes of higher energy particles that result from solar flares and CMEs.

SOLAR PLASMA AND ENERGETIC PARTICLES THAT IMPINGE UPON THE EARTH

SOURCE	ORIGIN	TYPES OF PARTICLES
Solar Wind	Open field regions in the corona	Mostly protons and electrons
CMES	Closed field regions in the solar corona	Protons, electrons and ions
Solar Flares	Solar corona and chromosphere	Protons, electrons and ions
Galactic Cosmic Rays	Supernovae, neutron stars and other galactic and extragalactic sources	Mostly protons, some ions and electrons

Particles Borne Outward in CMEs

Ejections of mass from the corona, called CMEs, are triggered by sudden realignments and reconnections in the magnetic fields that give form to this ethereal outer extension of the Sun. In the course of these violent events a long coronal streamer of the sort that extend outward for millions of miles from the surface of the Sun can be disrupted and rearranged, or wholly torn away. Chromospheric prominences—which are often found within the bulbous base of a coronal streamer and magnetically supported by it—are pulled up and flung outward from the Sun in the same catastrophic disruption.

The expulsion of a CME can often be traced to differences in the rates of solar rotation of adjoining latitude bands on the surface of the Sun. If one foot point of a towering magnetic arch that supports the bulbous base of a coronal streamer moves relative to the other one, it can twist the magnetic structure on which the streamer is built, and sever its moorings to the Sun. The emergence of new magnetic flux in the vicinity of a coronal streamer is another likely trigger of CMEs.

By any measure, the dimensions of a CME are truly immense, and as they move outward they continue to expand.

☉ ☉ ☉

The expulsion of a CME from the Sun might correspond on Earth to suddenly pulling up an entire forest of giant sequoias by their roots, and flinging them outward like a handful of weeds into space. Or perhaps the sudden volcanic obliteration of an entire island—as indeed happened off the tip of Sumatra one summer day in 1883.

But neither simile is apt, for in the corona the disturbance is of continental scale and often results in the violent obliteration of a sizeable piece of the entire outer corona. What we chance to see from the Earth in these minutes of unbridled solar rage are gigantic blobs of plasma, larger than the Sun itself, hurled outward from the star at velocities of up to 1000 miles per second: an almost unimaginable speed for so large an object, and fast enough to carry it from LAX to JFK in but three seconds' time. When the direction of an expelled CME carries it toward the Earth, we can expect it to reach the outer boundary of our magnetosphere in typically three to four days, although the fastest get here in but 19 hours.

CMEs are expanding pieces of solar plasma containing protons, electrons, and ions of different elements that are loosely bound together by the embedded magnetic field that they have taken with them from the Sun. As noted earlier, CME plasma is not particularly energetic, with energies per particle of but a few keV, which are not unlike those encountered in the solar wind. But higher speed CMEs can instigate shock waves in the solar wind through which they pass, accelerating ambient wind and suprathermal particles to energies measured in MeV. These high-energy, shock-driven particles generally run well ahead of the CME that was responsible for them, and are the principal cause of significant particle impacts at the Earth.

To be fully warned of their impending arrival, we need to have observed the disruption as it happens on the Sun, which is best secured from satellite-borne telescopes and coronagraphs. To tell whether it will strike the Earth we also need to know its initial direction, bulk speed (as distinct from the speeds of individual particles) and projected course. To forecast the likely consequences at the Earth we need to know more about the CME, including its likely impact on the slower-moving solar wind ahead of it and the polarity (north pointing or south pointing) of the magnetic field in the approaching plasma.

Particles From Solar Flares

Explosive solar flares also release streams of highly energetic particles outward toward the planets and into the heliosphere, although they are not the source of the largest impacts at the Earth.

Like CMEs, solar flares represent the instantaneous conversion and release of some of the magnetic energy contained in highly localized and magnetically unstable regions on the Sun, and from the acceleration of these charged particles in their passage through the outer solar atmosphere. Unlike CMEs, however, much of the truly immense magnetic energy that is converted in a flare is released as a short and intense burst of x-ray and EUV radiation.

Were our eyes able to see the disk of the Sun in the EUV and x-ray portion of the solar spectrum we would need no telescope to tell us when and where on the Sun a major flare occurred, which is not the case for visible radiation. As noted earlier, for a few minutes at the start of a solar flare the localized brightening in that one small area—seen in the light of x-ray and EUV radiation—would appear so intense that it would far surpass the brightness of all the rest of the solar disk.

Most of the atomic particles that are thrown off from a flare are electrons that were separated from atoms of hydrogen, helium, iron and other chemical elements in the hot chromosphere and corona. With them, though fewer in number, are the ions that remained when the electrons were freed.

Of all flare particles, solar energetic protons (or *SEPs*) pose the greatest hazard to both life and equipment in space, due to the combination of their mass and speed. The fastest stream outward from the star at half the speed of light, with energies per particle of tens of billions of electron volts. At these near-relativistic speeds they can travel the 93 million miles to the Earth in about fifteen minutes—about the time it might take us to eat breakfast.

Once within the Earth's atmosphere, SEPs will spend most of their prodigious energy in repeated collisions with atoms and molecules of air. But before that happens they can pierce almost any metallic shield.

Manned space flights are particularly at risk, and especially those missions that venture beyond the denser atmosphere and the bounds of the magnetosphere, whether on a short trip to the Moon or a much longer journey to Mars, and whether the crew is inside or outside the spacecraft. Nor would places of apparent shade on the surface of these distant landing sites offer certain protection, because the Sun throws curve balls.

☉ ☉ ☉

Upon leaving it, charged particles closely follow extended magnetic field lines that are curved into spiral form by solar rotation. At the Moon's distance from the Sun (the same as that of the Earth) the angle of arrival is tilted about 45° from the line of sight to the Sun, and on more distant Mars, about 55°. Moreover, because of the way incoming charged particles are deflected or "scattered" in all directions by the intervening medium, on the Moon or Mars as elsewhere in space there are as many solar particles arriving from the anti-solar direction as from the opposite, Sun-facing direction.

Given these circumstances, finding shelter may not be easy. Moreover, to take full advantage of the few available minutes of warning time, observers on the Earth must have witnessed the initial burst of the flare, recognized its severity and reacted quickly to sound the alarm.

The best hopes for extending the warning time for these major solar events are through more accurate predictions, issued as far in advance as possible, of when and where on the Sun a CME or major flare is likely to occur; or to avoid all manned spaceflight beyond the magnetosphere during the five or six years of maximum activity in the 11-year solar activity cycle, which is when the most CMEs and largest flares are most likely to occur.

Galactic cosmic rays are a horse of a different color. While there is hope of minimizing the effects of solar high-energy particle events through a combination of shielding and forewarning, galactic cosmic rays—due to both their extreme energies and omnipresence—present a far more difficult and as yet unsolved problem. In terms of cosmic ray exposure, the opposite option of scheduling long-duration space flights during times of maximum solar activity, when the flux of galactic cosmic rays in the inner heliosphere is significantly reduced, may prove to be the better gamble.

The Solar Wind Plasma

The charged atomic particles that are carried away from the Sun in slow or fast streams of solar wind plasma carry energies per particle of only about one keV. But they flow outward from the star every minute of every day and through all seasons of the 11-year solar activity cycle, and are often accelerated en route to the Earth.

Protons and other heavy particles in the solar wind can be accelerated by faster moving CMEs to energies that can exceed those in flare-produced solar energetic protons. Moreover, the threat of these CME-accelerated particles to space travelers or space equipment is greater than that from solar flares, since they last so much longer. Exposure to particles from flares extends typically from a few minutes to half an hour, while an accelerated stream of solar wind plasma can continue for a matter of days.

Particles in the solar wind plasma come from wherever there are open field lines in the Sun's corona: from the poles, the equator, and places in between, and in so doing spray outward from the Sun in virtually all directions.

Many that leave the polar regions of the Sun are channeled by magnetic lines of force that curve gracefully outward from the Sun toward lower solar latitudes. Many of these and all that come from more equatorial latitudes will continue outward into space on fixed, straight-line courses that take them into the realm of the planets. As they proceed outward, the speeding plasma streams—initially more compact—gradually expand into an ever-wider cone: like pellets from the muzzle of a shotgun.

Eventually, they will fill every part of the immense heliosphere. But in the first leg of their long outward race—as they pass the milestones that mark the orbits of first Mercury, then Venus, Earth, Mars, and the more remote planets—the individual streams remain largely separate and distinct: in spite of the pushing and shoving that happens when they are overtaken by faster moving particles in CMEs.

Only a miniscule fraction of what the Sun releases in the solar wind will encounter the Earth or other planets, all of which are all small targets at great distances. Seen from the Sun, the Earth would appear about as large as a ping pong ball held up at the far end of a football field: more like a star than a planet, and though perhaps as bright, only half as large as the planet Jupiter appears to us.

Characteristics of Slow Solar Wind Streams

Slow-speed streams of solar wind plasma, made up of predominantly protons and electrons and a much smaller number of heavier ions, move outward at about 200 miles per second. On average they will arrive at the orbit of the Earth in four to five days, and at Mars in about a week. As in faster solar wind streams, the energies of individual electrons fall in the range of one to twenty electron volts and for protons about 700 eV (0.7 keV) per particle. Apart from this is the energy in the bulk motion of the slow-speed solar wind plasma, which is derived from the average of individual particles.

The initial trajectory followed by slow-speed plasma is determined by the shape and placement of individual streamers. Solar plasma coming from streamers in or directed toward higher latitudes will expand outward in a direction that is tilted upward from the plane of the Earth and other planets. The flow of particles from a streamer whose base lies at lower solar latitudes, on or near the Sun's equator, will be roughly parallel to this plane. Since streamers are found at low latitudes in all seasons of the 11-year solar cycle, the Earth is continually bathed in the slow-speed solar wind, and less commonly in gusts of the high-speed wind.

High-Speed Solar Wind Streams

High-speed streams of plasma in the solar wind race outward from low in the corona at speeds of 350 to 550 miles per second that clearly distinguish them from those in the slower wind. Both high- and low-speed streams of plasma are present in the space that separates the Earth from the Sun, although they cannot interpenetrate, for they follow magnetic field lines from different regions on the Sun. As emphasized earlier, either slow or fast solar wind streams can be disturbed by suprathermal particles which are carried outward from the Sun in either flares or in CMEs. These can produce shock waves in the ambient wind through which they pass, greatly accelerating some of the plasma.

At these higher velocities the high-speed solar wind plasma arrives at the orbit of the Earth within 2 to 3 days and at Mars in 3 to 4. Energies per particle are also somewhat greater: typically ten to thirty electron volts per particle as opposed to a slow-speed average of about 10 eV. The bulk-flow properties of the high-speed wind, though quite different for protons and electrons, are in this case comparable to the thermal energies of the particles within it.

As noted earlier, high-speed streams can be traced to regions on the Sun—called coronal holes—where the density is much lower than in the surrounding corona and where there are no concentrated magnetic fields or coronal streamers. These regions of open magnetic field lines offer a direct path for the escape of high-speed solar wind plasma.

The coronal holes that persist the longest are found at the two poles of the Sun, where they are present in all but the most active years of the solar cycle. Other large coronal holes also appear in middle and lower latitudes on the Sun when it is less active, separating large magnetically-active regions on the disk. The result is a strong solar cycle effect in the solar latitudes where coronal holes appear, and in the number of expected high-speed streams in the solar wind.

☉   ☉   ☉

When the Sun is totally eclipsed and its poles are sufficiently clear of high-latitude coronal streamers—as in years of minimum activity—we can actually see the high-speed streams of plasma that flow radially outward from the poles. They appear as a crown of slender white coronal plumes, made visible to us by the same process that illuminates all the rest of the corona: namely, the scattering or redirection of white light from the photosphere by coronal electrons.

Sectors in the Sun's Extended Magnetic Field

The open magnetic field lines that are carried in the solar wind plasma are drawn outward from the star like trailing ribbons with one end still attached to the Sun. Plasma streams move outward from the Sun along radial lines. But because they are tethered to the star, the magnetic field lines that trail behind them are wound around the Sun in loose spirals by solar rotation, like the spiral arms of a rotating galaxy.

The magnetic signature carried outward in each expanding stream is impressed upon it at its place of origin on the Sun. As a result, the magnetic properties of the expanding solar wind at low latitudes are divided into discrete segments of common polarity called sectors. Some have originated in regions of predominantly "outward pointing" polarity on the Sun, and others in places of predominantly "inward pointing."

Since these sources are commonly traced to adjoining regions of opposite magnetic polarity, the solar wind streams emanating from them expand outward in adjoining, oppositely polarized sectors. As these curved sectors, fixed at the Sun and rotating with it, sweep by the Earth—like the sprays from a revolving lawn sprinkler—the polarity of the interplanetary magnetic field sensed at the Earth will switch from one polarity to the other each time a sector boundary sweeps by.

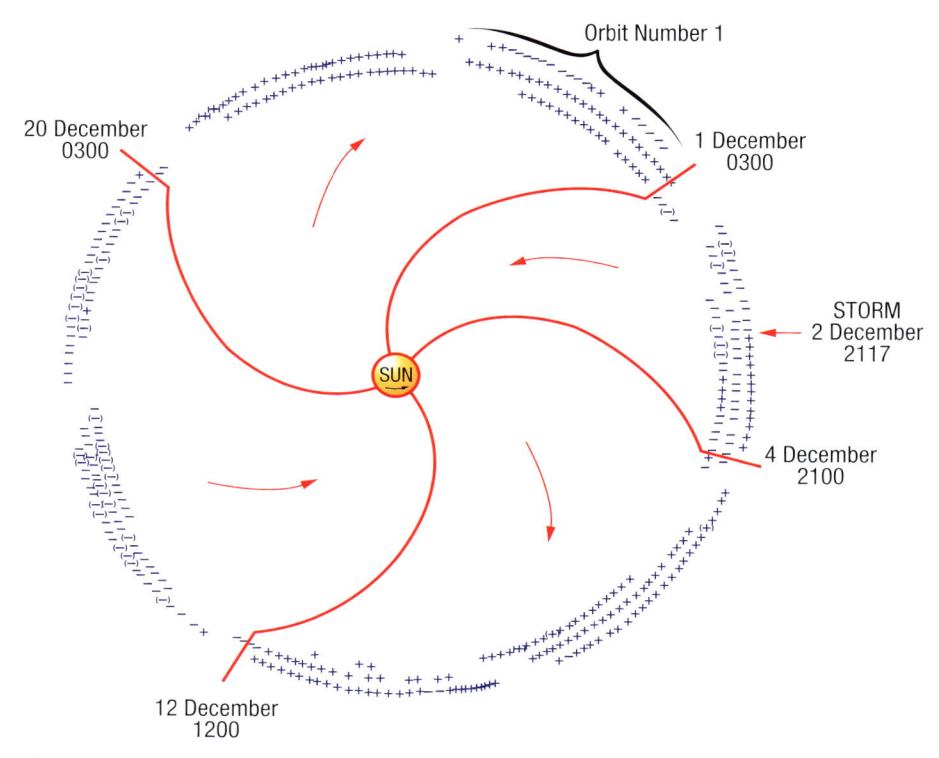

A schematic representation of the Sun, rotating counterclockwise as seen from above its north rotational pole, and the magnetic characteristics of the solar wind that flows outward from it. Particles coming from extended regions of prevailing positive or negative magnetic polarity on its surface carry the local dominant polarity with them, forming magnetic sectors of like sign, four of which are shown in this example. This causes the magnetic polarity of the wind received at the orbit of the Earth to switch from prevailing positive (indicated here from actual dated measurement by blue + signs with red outward directed arrows and blue negative -, with red inward directed arrows) as the boundaries of the sectors are swept by the Sun's rotation past the Earth. The curved form of the sectors and of the tethered magnetic field lines carried within them results from the combination of solar rotation and the radial outward flow of the solar wind.

By the time the solar wind plasma reaches the orbit of Venus—about ¾ as far from the Sun as we—the rotation-induced curvature in the sectors has turned the direction of their original magnetic polarity through about 30°. Farther on, at the orbit of the Earth the angle is more nearly 45°.

Although the plasma particles are carried radially outward from the Sun, the location from which they came is carried by solar rotation east-to-west (left to right) across the face of the Sun during the time of their outward travel. Due to this effect, streams of plasma emitted from solar features that are near the apparent central meridian of the solar disk will curve away from the Sun-Earth line as they move outward, and will ultimately miss the Earth. It is not until the rotation of the Sun has carried an active region several days westward—to an

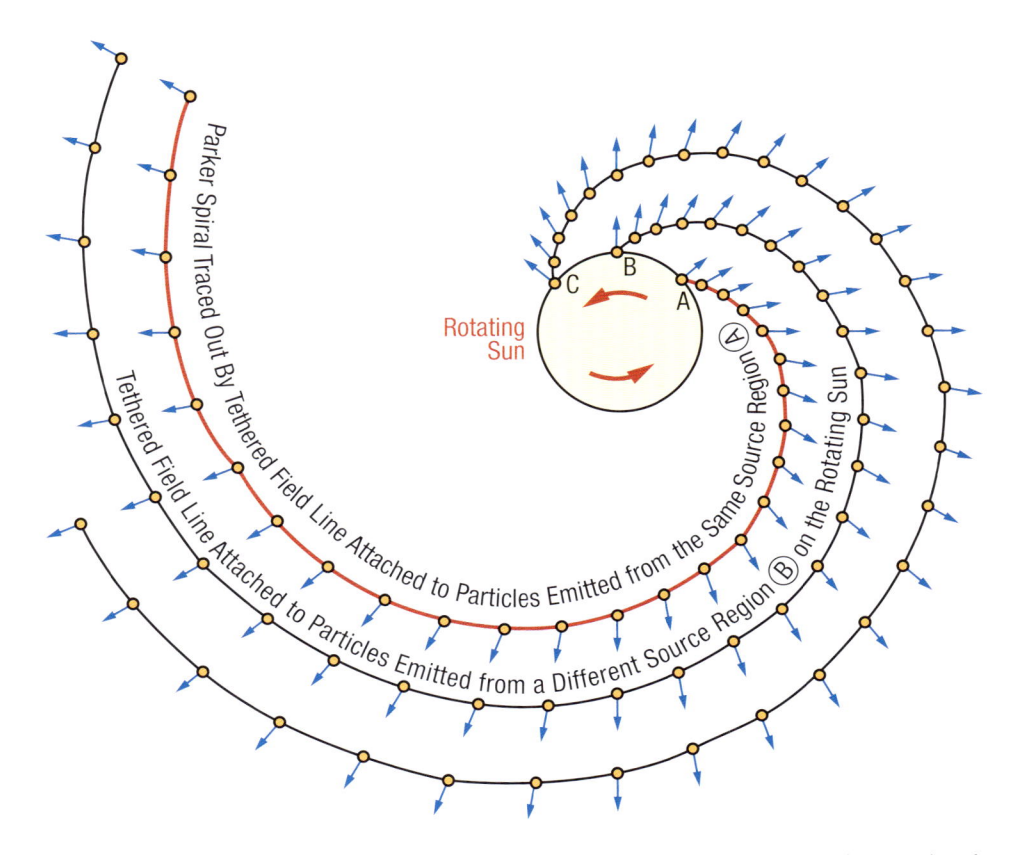

The curving magnetic field lines, tethered at the rotating solar surface, which are carried outward in the solar wind from three source regions (A, B, or C in this schematic drawing), generate the abiding heliacal form of the Sun's extended magnetic field. Blue arrows indicate the always-radial direction of the wind front as it progressed outward.

apparent solar longitude of about 45° west—that it is in a position from which plasma expelled from it is most likely to strike our planet.

Pushing and Shoving on the Way to the Earth

Outward moving plasma in fast- and slow-speed solar wind streams can interact mutually, slowing the fast wind and speeding up the slow. But as noted earlier the more telling disruptions on their way to the Earth occur when these plasma streams in the solar wind are overtaken by more energetic particles from CMEs and less often, from solar flares.

In these cases the CME or flare particles, which move much faster than the overtaken plasma, produce transient disturbances in the solar wind that, in turn, can provoke large geomagnetic storms on the Earth, with impacts on life and society.

A typical disturbance of this kind will abruptly raise the bulk flow speed in the solar wind plasma from perhaps 200 to 600 or more miles per second. This produces a region of high pressure on the leading edge of the disturbance that is bounded by a forward-moving shock wave that compresses and heats the solar wind plasma. Shock-acceleration in the solar wind plasma produces flows of very high energy particles which are responsible for the longest-duration and terrestrially most significant SEP events.

When Solar Particles Strike the Earth

Streams of solar wind plasma that come upon the Earth are initially deflected by pressure forces at the magnetospheric bow shock and soon after, at the magnetosphere itself. Nevertheless, a number of these fleeing particles from the Sun—depending on the magnetic polarity and energy of the incoming plasma, its speed, and where it comes in contact with the magnetosphere—will still find their way through or around these barriers.

The passing hordes that are wholly deflected around the Earth still leave their mark on the planet by distorting and distending the shape of the magnetosphere, which ultimately affects conditions at the surface of the Earth.

To streams of approaching plasma, the Earth's magnetosphere presents an obstacle much like a protruding stone in the middle of a tumbling mountain stream.

When solar wind plasma moving at supersonic speed encounters an obstacle like the planet Earth it develops ahead of its path a dense and abrupt compression called a shock wave through which the solar plasma must pass. The same is true in a more familiar instance, when the object is moving and the fluid is fixed, as when—accompanied by an audible sonic boom—an accelerating jet aircraft exceeds the speed of sound.

In the case of the solar wind and the Earth, the obstacle is not the solid planet itself, but the lines of force of its magnetosphere. To streams of charged atomic particles moving at supersonic speeds these present an equally disturbing obstacle, since the two magnetic fields—one from the Sun, the other at the Earth—cannot interpenetrate.

The shock wave produced at this encounter is called the bow shock of the magnetosphere, due to its similarity to the bow wave that forms in the water ahead of the bow a moving ship. In the nautical case, the wave is formed not where the sharp edge of the bow slices the water but some distance ahead of it.

In the same way, the bow shock in the solar wind is created well ahead of the outer boundary of the magnetosphere and at a great distance from the Earth itself. Depending upon the speed and energy of the approaching solar wind, the bow shock is met typically 50,000 miles ahead of the Sun-facing side of the Earth, and sometimes as far as 80,000 miles, or as much as a third of the distance to the Moon.

Solar Wind Plasma   Bow Shock   Magnetopause

In passing through the curved bow shock—a zone of turbulence often less than 100 miles thick—the onrushing solar plasma is heated, slowed to sonic speeds and deflected.

Bow Shock   Magnetopause

The next obstacle it will encounter is the magnetopause, the shell or outer boundary of the Earth's magnetic field, which lies roughly half-way between the bow shock and the solid Earth. Even there, the streaming plasma still has about 40,000 miles yet to go to reach the planet itself.

Bow Shock

As it approaches this second barricade, the onrushing plasma compresses the curved lines of force that connect the two magnetic poles of the planet, flattening the shape of the Sun-facing side of the magnetosphere and pushing the magnetopause closer to the surface of the planet it protects.

Magnetopause

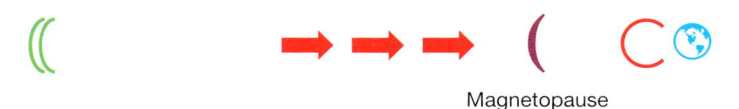

Because the Earth's magnetic field is so much stronger than that carried in the oncoming solar plasma, it repels the brunt of the surge. Under normal conditions the deflected solar plasma is channeled around the magnetopause to continue its course, like a diverted stream, on beyond the Earth.

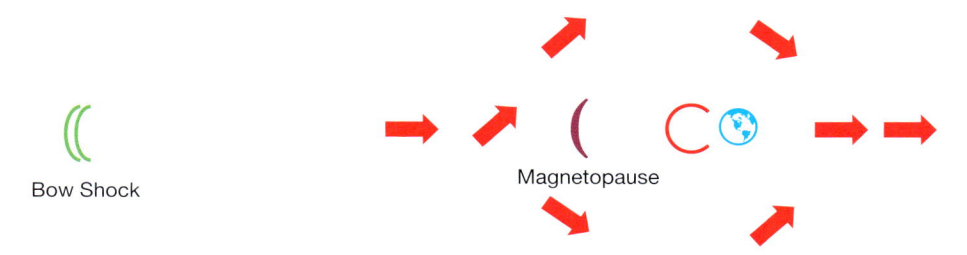

Bow Shock Magnetopause

Through the Guarded Gates

However, should the magnetic polarity in the stream of incoming solar particles be oriented such that a significant component is directed opposite to that of the Earth—that is, south-pointing instead of north-pointing—the transported solar field and that of the Earth can directly connect to each other. This process of magnetic reconnection on the day-side of the magnetosphere opens the normally closed magnetopause, allowing the direct entry of charged particles from the Sun into the Earth's magnetosphere and on down into the thermosphere.

The reconnection process creates open field lines, rooted at one end near the poles of the Earth, that thread the magnetosphere, affording paths of access for the newly admitted particles. In so doing, they temporarily expand the diameter of the polar cusps, enlarging the funnel-shaped openings at the Earth's magnetic poles, thereby increasing the area through which diverted solar plasma moving through the magnetosheath can flow into the magnetosphere.

The momentum of the solar wind drags the open field lines rooted at the two poles of the Earth downwind into a long cylindrical magnetotail, made up of a north lobe and a south lobe, each D-shaped in cross section, and stacked back-to-back ⊖. Since the field lines in one originate at the north magnetic pole of the Earth and the other at the south, the lobes have opposite magnetic polarity, and are separated like a sub sandwich by a thin slice of hot plasma and weak magnetic field called a plasma sheet.

Closed Magnetosphere

Magnetopause

Solar Wind

Open Magnetosphere

Magnetopause

Solar Wind

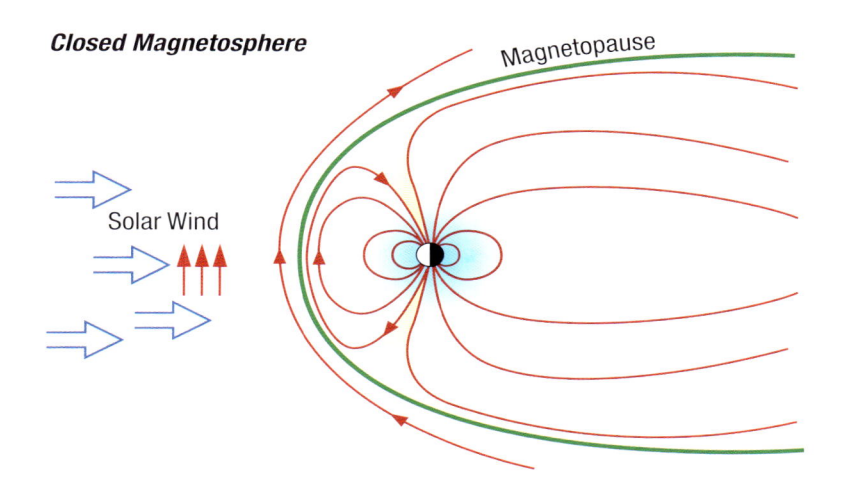

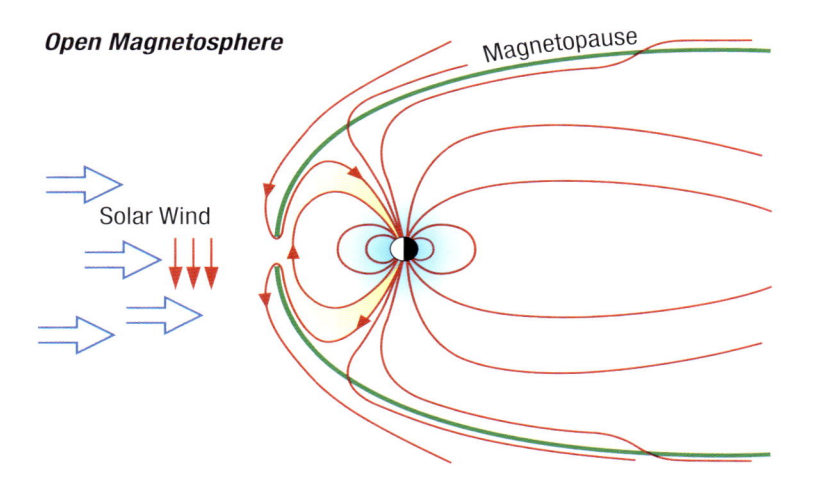

Upper right: Lines of force of the Earth's magnetic field, portraying with arrows their presently-fixed, north-pointing direction of polarization. Field lines emanating from either magnetic pole connect with the opposite pole, defining a closed magnetic field, as shown here. The field lines streaming out to the right, "downwind" side in this figure are also closed, connecting with lines from the opposite pole but at a distant point well beyond the limited portion of the Earth's field shown here. At upper left: the approaching solar wind that flattens the Earth's field on the Sun-facing side. In this case it carries the imprint of a similarly north-pointing magnetic field (red arrows) from the Sun's surface. Since it is parallel to that of the Earth no Sun-Earth connection is made.

Lower left: An approaching solar wind or other disturbances such as a CME that carries instead a south-pointing magnetic signature. Lower right: the effect of this oppositely-directed field on the Earth's magnetosphere. Within a limited point near the Earth's equator on its Sun-facing side, the direction of the solar field is exactly opposite to that of the Earth, allowing a direct electric connection between the two, through a process called magnetic reconnection. For a short time, this opens a gateway for the transfer of charged particles from the Sun directly into the magnetosphere, where within the yellow region, they follow terrestrial field lines into the upper atmosphere of the planet to produce a magnetic storm with accompanying aurorae and potentially major disruptions at the Earth's surface.

In the distant magnetotail, stretched-out field lines attached to the Earth's north magnetic pole are pinched by the force of the streaming plasma into contact with those of the opposite magnetic polarity rooted in the south magnetic pole. In these regions of contact—called neutral lines—solar plasma that has streamed outward in the magnetosheath is able to enter the outer magnetosphere and become entrapped. Once there it is channeled all the way back to the Earth and into the inner magnetosphere and radiation belts.

There is a final means of entry into the magnetosphere, that is also highly restricted: in this case for cosmic and solar particles (as opposed to plasma streams) that are sufficiently energetic to breach the protective magnetopause. When this happens these highly energetic particles make their way directly into the upper and middle atmosphere, where they are ultimately stopped by collisions with atoms and molecules of air. To qualify for brute force entry, charged particles must carry very high energies—in the range of about 500 MeV or more—which include both galactic cosmic rays and the most energetic solar particles, such as solar energetic protons (SEPs). How deeply they penetrate into the atmosphere is a function of their energy, their angle of approach and the magnetic latitude at which they enter.

Magnetic Reconnection

As noted above, when the normally-closed magnetopause is for a time opened by magnetic reconnection, it creates a network of open-ended field lines, with one end rooted in the area of the Earth's polar caps. This temporary opening of the magnetopause allows the direct entry of solar particles into the low-latitude, Sun-facing magnetosphere, as well as solar particles that have been deflected around the Earth in the magnetosheath.

Those that enter the magnetopause at the site of magnetic reconnection are directed along the open field lines toward one of the magnetic poles, where some are funneled directly into the upper atmosphere through the polar cusps. Some of the particles traveling around the outer boundary of the magnetosheath also find their way through the polar cusps and into the atmosphere.

Others within the magnetosheath follow newly-opened field lines in the magnetotail, where at one of several possible neutral lines they execute a U-turn to head back in the direction from which they came, toward the Earth and into the inner magnetosphere.

A near-Earth neutral line where this can happen is temporarily established during magnetic reconnection at a distance of about 25 R_E or 100 thousand miles downwind from the planet. A more permanent and more heavily used distant neutral line is located more than 300 thousand miles farther out in the magnetotail, which puts it well beyond the orbit of the Moon.

The opening of the magnetopause during magnetic reconnection events also allows an exchange of particles in the opposite direction: out of the magnetosphere, through the magnetopause and into the magnetosheath where they are returned, unused, back into interplanetary space.

During times of magnetic reconnection the magnetosphere extracts about 2% of the incident energy of the solar wind, of which more than half is returned, unused, to space. But even the small fraction that is retained—stored in the magnetotail in the form of magnetic field energy—can have major impacts. The connection can be maintained for hours or days, and during this time the effects of the direct connection are felt throughout the length and breadth of the entire magnetosphere and in the upper atmosphere of the Earth as well, with significant effects on a range of human activities.

⊙ ⊙ ⊙

The orientation of the magnetic field borne in an approaching stream of solar plasma can lie anywhere from opposite to that of the Earth's field to completely parallel to it, or any angle between these two extremes. As previously noted, magnetic reconnection between the approaching solar field and that of the Earth is possible when the fields are oriented oppositely, or more precisely, when a significant component of the vector that describes the orientation of the approaching solar field is exactly opposite to that of the Earth's fixed field.

One of the two simplest cases can be envisioned as the coming together of the shafts of two arrows ↓↑ that are *anti-parallel*, or pointed in exactly opposite directions. Under these conditions, or in the more likely case in which a significant component of the solar field vector is anti-parallel to that of the Earth, a direct connection is possible. But when the orientation of the magnetic field in the incoming plasma stream is the same or largely the same as the portion of the Earth's field with which it comes in contact—i.e., with our schematic arrows pointed in the same direction ↑↑ —a meaningful connection is not possible.

⊙ ⊙ ⊙

The present orientation of the Earth's magnetic field is by convention described as "north pointing" in the sense that the direction of magnetic force at the equator is from the south to the north magnetic pole ↑C🌏 .

And so it has been, with no drastic change in polarity, since the time of the last reversal, when early Stone Age people began to wander into what is now Europe: three-quarters of a million years ago.

When the Earth's magnetic field next reverses direction, from north-pointing ↑C🌏 to south-pointing ↓C🌏 , the opposite will apply.

Magnetic reconnection with incoming solar plasma will then take place only when a component of the incoming solar field is oriented in the opposite "north pointing" direction, although the change, once accomplished, will make no difference in the frequency of magnetic reconnection events. Were we here at that time, we would notice not the switch in polarity but a protracted period between states: a period of magnetic confusion when the strength of the Earth's protective field will fall almost but not quite to zero.

☉ ☉ ☉

The more-superficial surface magnetic field of the Sun also switches polarity, though at a much faster rate, following a solar magnetic cycle that is twice the length of the 11-year activity cycle. Toward the end of each 11-year cycle the north and south magnetic poles of the Sun switch polarities, from "north pointing" to "south pointing" or vice versa, and remain that way until the end of that cycle and the beginning of the next, when they will switch again, completing one 22-year solar magnetic cycle. Thus the Sun, so often invoked as a symbol of stability and constancy, is the more skittish partner in the Sun-Earth system, having undergone more than 68,000 magnetic reversals since the last time the Earth changed its magnetic mind, about 750,000 years ago.

The magnetic polarity carried in streams of plasma that reach the Earth is determined not simply by the Sun's 22-year magnetic cycle but by the polarity of the local magnetic field in the part of an active region from which the plasma originates, which can be either "north-pointing" or "south-pointing" at any phase of the Sun's 11- or 22-year cycles. Thus the magnetic orientation of the heliospheric magnetic field sensed at the Earth is a continually varying parameter, as a consequence of the presence of discrete regions of opposite polarity on the surface of the Sun and its 27-day rotation.

SOME DISTANCES EXPRESSED IN UNITS OF THE RADIUS OF THE EARTH

FROM THE CENTER OF THE EARTH TO:	DISTANCE IN R_E [≈4000 MILES]
Mean Sea Level (0 MILES)	1
Peak of Mount Everest (5.5 MILES)	1.0014
Auroral Displays (260 – 580 MILES)	1.05-1.08
Near-Earth Orbit	1.10
Geosynchronous Orbit	6.6
Magnetopause in the Direction of the Sun	10
Magnetopause above the N and S Poles	13
Bow Shock at the Nose of the Magnetosphere	15
Nearest Neutral Line in the Magnetotail	25
The Moon at Mean Distance	60
Farthest Neutral Line in the Magnetotail	≈ 100
End of the Magnetotail	> 1000
Venus at Closest Approach	6,400
Mars at Closest Approach	12,200
The SUN	23,500

Effects of Changes in the Earth's Magnetic Field

The degree to which the magnetosphere can deflect incoming solar and cosmic particles depends on its strength, the energies of the incoming particle, and their direction of approach relative to the geometry of the Earth's field lines.

Mercury's magnetic field, for example, is about 100 times weaker than our own. Because of this and the planet's proximity to the Sun—more than twice as close as we—its magnetic field is very highly compressed on the Sun-facing side. As a result, it can hold off the pressure of solar wind plasma no farther than about 1500 miles: a distance above the surface of the planet that is about equal to its radius. This and Mercury's lack of an atmosphere—which allows a direct interaction between its magnetosphere and the solid surface of the planet—permit easier access to incoming energetic particles.

The Earth's field, on the other hand, is sufficiently strong and oriented in such a way that most incoming particles are fended off—at a distance on the Sun-facing side of about 40,000 miles above the solid surface. But it hasn't always been that way, nor will it always be, for the strength and orientation of the Earth's magnetic field is always changing.

The locations of the Earth's north and south magnetic poles wander as time goes by, seldom if ever coinciding with the locations of the two rotational poles. In the Arctic, the peripatetic north magnetic pole is today meandering along near Melville Island in Canada's Northwest Territories, some 800 miles south of the North rotational pole: about the distance that separates Chicago from New Orleans. And it continues to move at a speed of about one mile per day, in response to changes in the internal dipole field of the planet.

⊙ ⊙ ⊙

Two corrections need be applied to correct the difference between what a magnetic compass indicates as "north" and "true north" as defined by the direction to the Earth's rotational pole. The first of these corrects for the difference between the positions of the north magnetic and north rotational poles and is called the variation or magnetic declination: a correction that navigators and surveyors routinely apply when using a magnetic compass. (The other, called the deviation, is a local effect that arises from the presence of magnetized metal or ore in the immediate vicinity of the compass itself.)

Since the variation changes from year to year and place to place with the meandering of the poles, the correction for variation indicated on maps, navigational charts and flight maps require frequent updating. Because aircraft commonly employ magnetic compasses, the large Arabic numbers that are painted at each end of airport runways (e.g., 12 to indicate a magnetic heading of 120°, or 30 for 300°) must also be corrected and repainted from time to time.

⊙ ⊙ ⊙

A more significant change—in terms of its effect on the efficacy of our magnetic shield—is a well-established long-period variation in the *strength* of the Earth's field.

Five hundred years ago, when wooden ships set sail from ports in the Old World to explore the New, the strength of the magnetic field that kept their compasses pointed north was about 15% stronger than it is today. And stronger still 1300 years before that when, at about the time the Roman Empire reached its greatest extent, the Earth's magnetic field—after a long, three thousand year climb—reached a peak strength that was almost 40% greater than today.

More dramatic than these several-thousand year variations in the strength of the field are longer-spaced reversals in the polarities of the Earth's two magnetic poles: at which time the north and south magnetic poles exchange polarity, though not exactly at the same time.

In the course of these slow and imperfectly synchronized events, the effective strength of the Earth's field—and with it, the protection afforded by our magnetic shield—drops precipitously low, if not to zero, for long periods of time.

Repeated occurrences of magnetic reversals have been identified in paleomagnetic records of the last 80 million years, separated by periods of fixed direction that persist for about 100 thousand to almost one million years. Although the recurrence and timing of these events is highly irregular, the next reversal is widely conceded to be overdue, since the last one occurred in the early Pleistocene epoch, about 750,000 years ago.

Cosmic Rays

The Earth is continually bombarded with charged atomic particles not only from the Sun but from violent events—such as the eruption of a supernova—that happen much farther away, either within our immense galaxy, the Milky Way, or far beyond it. They are accelerated in the shock waves that inevitably follow such cosmic explosions, and are known as *cosmic* rays, since at one time all were thought to come from outside the solar system, in the broader universe, or cosmos. More than 100 cosmic rays speed through every square foot of our upper atmosphere each second of every day.

It was more recently found that while some, more properly called galactic cosmic rays or GCRs, do indeed come directly to us from far away, another class (initially called anomalous cosmic rays, or ACRs, and later, "pick-up ions") are born within the heliosphere.

ACRs are created when neutral atoms from the interstellar medium flow into the outer heliosphere. Those that penetrate into the near-Sun vicinity are ionized by solar short-wave radiation, and as charged particles are picked up by the magnetic field carried in the solar wind and accelerated to high speeds in passing through the termination shock. Some of these reach the Earth. There, they can be distinguished from other, galactic cosmic ray particles that reach the Earth by both lower particle energies and lower charge. ACRs are singly-charged ions with but one missing electron; GCRs, in contrast, have typically been stripped of all of their orbital electrons and carry a much higher positive charge.

Cosmic rays, whether ACRs or GCRs, are not "rays" at all—in the sense of rays of electromagnetic radiation in visible, ultraviolet, or infrared wavelengths—but ions: the same sorts of atomic particles that the Sun sends our way, all day, every day in the solar wind, and more sporadically from flares and CMEs. About 90% of all cosmic rays are *protons*; roughly 9% are the positively-charged nuclei of helium atoms (called alpha particles); and the remainder electrons and the charged nuclei of heavier elements. Most GCRs can be readily distinguished from particles of solar origin in that they are far more energetic.

GCRs that reach the near-Earth vicinity come in a wide range of energies. In passing into and through the heliosphere only the most energetic—those with energies greater than a few hundred MeV—are unaffected by heliospheric processes that slow them down, reducing the energy of a GCR by several hundred MeV during its passage through the heliosphere to the orbit of the Earth. Thus, ordinary GCRs that are detected at the Earth—whether coming from within our galaxy or beyond it—must have arrived at the boundary of the heliosphere with higher energies than what we observe here.

By the same token, GCRs that come from beyond our galaxy must have started their long journey with even greater energies. In order to slip through the combined magnetic field of the stars and other objects in our galaxy these intruders from truly distant worlds must begin their travels with energies in the range of about 10^{20} electron volts per particle (100 million trillion electron volts!). To these super-powered particles, breaching what are to them the relatively weak defenses of the heliosphere, and soon after that, our even weaker magnetosphere, must seem like a piece of cake.

THE PRICE OF ADMISSION IN ELECTRON VOLTS FOR CHARGED ATOMIC PARTICLES

PURPOSE	MINIMUM ENERGY REQUIRED, PER PARTICLE	IN GIGA (10^9) ELECTRON VOLTS
Entry Into the Galaxy	10^{20}	10^{11} GeV
Passage Into the Inner Heliosphere	3×10^8	.3 GeV
Unrestricted Entry Into the Magnetosphere • At the poles • At mid latitudes • At the equator	GRATIS 10^8 10^{10}	1 GeV 10 GeV

The Fate of Cosmic Rays

The microscopic examination of lunar rocks brought back to Earth by Apollo astronauts has shown that through at least the last 100 million years (since about the middle of the Cretaceous period of Earth history) the Moon has been bombarded by high energy cosmic particles at about the same rate as recorded on the Earth by neutron monitors today.

Since our close neighbor the Moon receives—per square foot—the same number of cosmic rays that strike the top of the Earth's atmosphere, this lunar finding of past cosmic ray flux applies to the Earth as well. But with no magnetosphere or atmosphere to buffer it, the Moon's long pummeled surface takes the full force of every solar or cosmic punch.

It is the Earth's atmosphere and not the magnetosphere that shields the surface of the Earth from the most energetic cosmic rays. Their energies are absorbed in collisions with atoms and molecules in the Earth's coating of protective air, much as an arrow, or a bullet, shot into water soon loses its thrust. Those cosmic rays that do make it to the surface are limited to the weaker, secondary particles that were created near the end of a downward cascade of collisions that started in the upper atmosphere.

Taking most of the hits are atoms and molecules of nitrogen and oxygen: the most abundant constituents of the atmosphere, and these encounters are not inconsequential. The collision of a speeding, high-energy cosmic ray with a more lethargic atom or molecule is like the crash of a high-velocity bullet against a crystal goblet.

In the course of each of these violent impacts, the impacted atoms of air are shattered and torn asunder: producing a second generation of less energetic subatomic particles, each of which carries away a portion of the energy imparted by its parent cosmic ray. The heaviest and most energetic of the litter are protons and neutrons. Electrons are also produced, as are short-lived muons and pions that live for no more than a few microseconds.

These "daughter" or secondary cosmic ray particles extend the chain of collisions downward into the middle atmosphere, and ultimately a few make it to the ground. On the way down protons and neutrons soon suffer further collisions with other atoms of air, resulting in a <u>third</u> generation of sub-atomic particles. The protons and neutrons among these will in like manner generate a <u>fourth</u> generation; and so on, until all of the energy brought into the atmosphere by the parent primary cosmic ray has been spent: most of it—as with other families well endowed—by heirs.

This process of energy depletion through repeated collisions and divisions produces what is called a cosmic ray shower or cascade. They are initiated in the upper atmosphere and through repeated cascades reach down into deeper and denser layers of the atmosphere until the number of subatomic particles that are produced reaches a maximum and then begins to decrease.

Maximum production is reached at an altitude of but ten or eleven miles above the surface of the Earth: the very height at which the *Concorde*—like a daredevil leaping through a wall of flame—made its fleeting trips. Below that, as the atmosphere thickens, the generations of cosmic rays finally lose all their punch, but barely in time to spare us and all the other bottom feeders in the ocean of air in which we live.

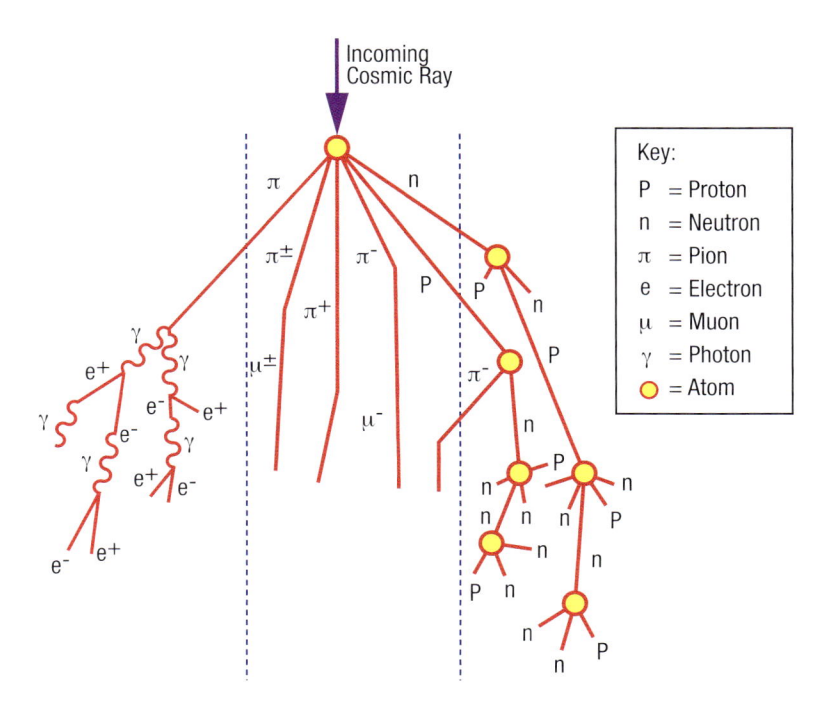

The shower of secondary atomic particles and radiation produced by the collision of a single incoming cosmic ray (top, center) with an atom of oxygen or nitrogen (first yellow circle) in the upper atmosphere, ultimately distributing the cosmic ray's not inconsiderable energy among less energetic, stable and unstable nuclear atomic particles, radiation in photons, and electrons. Here the initial impact releases (right panel) a proton and neutron, each of which initiates further cascades by colliding with other atoms of air (yellow circles.) Also released are less stable nuclear particles known as pions and muons (middle panel) as well as electrons and electromagnetic radiation (left panel.)

Could we see one of these cosmic ray cascades unfold—high overhead on a starry night—it would look much like the bursts and spreading trails of colored light that elicit oohs and aahs during aerial displays of fireworks on the 4th of July: although in this case each pyrotechnic display would last no longer than the wink of an eye.

The flux of cosmic ray and solar particles that reach the Earth behave oppositely and are very nearly mirror images of each other: when one is up the other is down. When the Sun is more magnetically active and spotted, the number of secondary cosmic ray neutrons that reach the Earth is reduced by about 20%. When the Sun is less active, more are recorded. A repeated pattern of 22 years, in step with the solar magnetic cycle, is also apparent in the Earth's receipt of cosmic rays as an alternation in the characteristic shape of successive 11-year cycles of cosmic ray flux, apparent in plots of cosmic rays vs time. In plots of cosmic ray incidence vs. time, 11-year cycles that reach a distinct peak alternate with cycles that are flat-topped, i.e. exhibit a prolonged maximum.

The last maximum in the 11-year solar cycle was reached in 2001. Since then, while the number of sunspots was gradually falling, year after year, the flux of cosmic rays—recorded continuously by neutron monitors at high-altitude stations in Colorado, Vermont, Antarctica and a host of other sites—reached a cosmic ray *maximum* (and a sunspot *minimum*) in 2008.

Sunspots themselves have no direct effect on cosmic rays, but disturbances in the flow of solar wind plasma do. For the most part, these can be traced to the interaction between fast-moving CME plasma and the ambient solar wind, which occur more often in years of maximum solar activity, since CMEs are more energetic and more frequent then. At these times of greater turbulence, incoming cosmic particles are more likely to be scattered and deterred from reaching the Earth. When the Sun is less active the opposite is true.

RELATIVE NUMBER OF HIGH-ENERGY ATOMIC PARTICLES THAT REACH THE EARTH'S MAGNETOSPHERE

SOURCE ↓	RECEIVED AT TIMES OF MAXIMUM SOLAR ACTIVITY	RECEIVED AT TIMES OF MINIMUM ACTIVITY
From the Sun	**More**	Fewer
From the Cosmos	Fewer	**More**

We can also expect the Earth's receipt of galactic cosmic rays to vary, periodically, on a much slower scale as the Sun makes its orbit around the center of the Galaxy: a circle so large that it takes almost 250 million years to complete a single lap around it.

The longest record of cosmic ray flux at the Earth is taken from the analysis cited earlier of lunar rocks brought back during the *Apollo* missions of the 1970s. From these rudimentary natural diaries one can infer that the flux of galactic cosmic rays—with enough energy to leave their tracks on the face of exposed lunar rocks—has not changed very much during the 100 million year period (a little less than half a turn around our Galaxy) that is sampled in the exposed material.

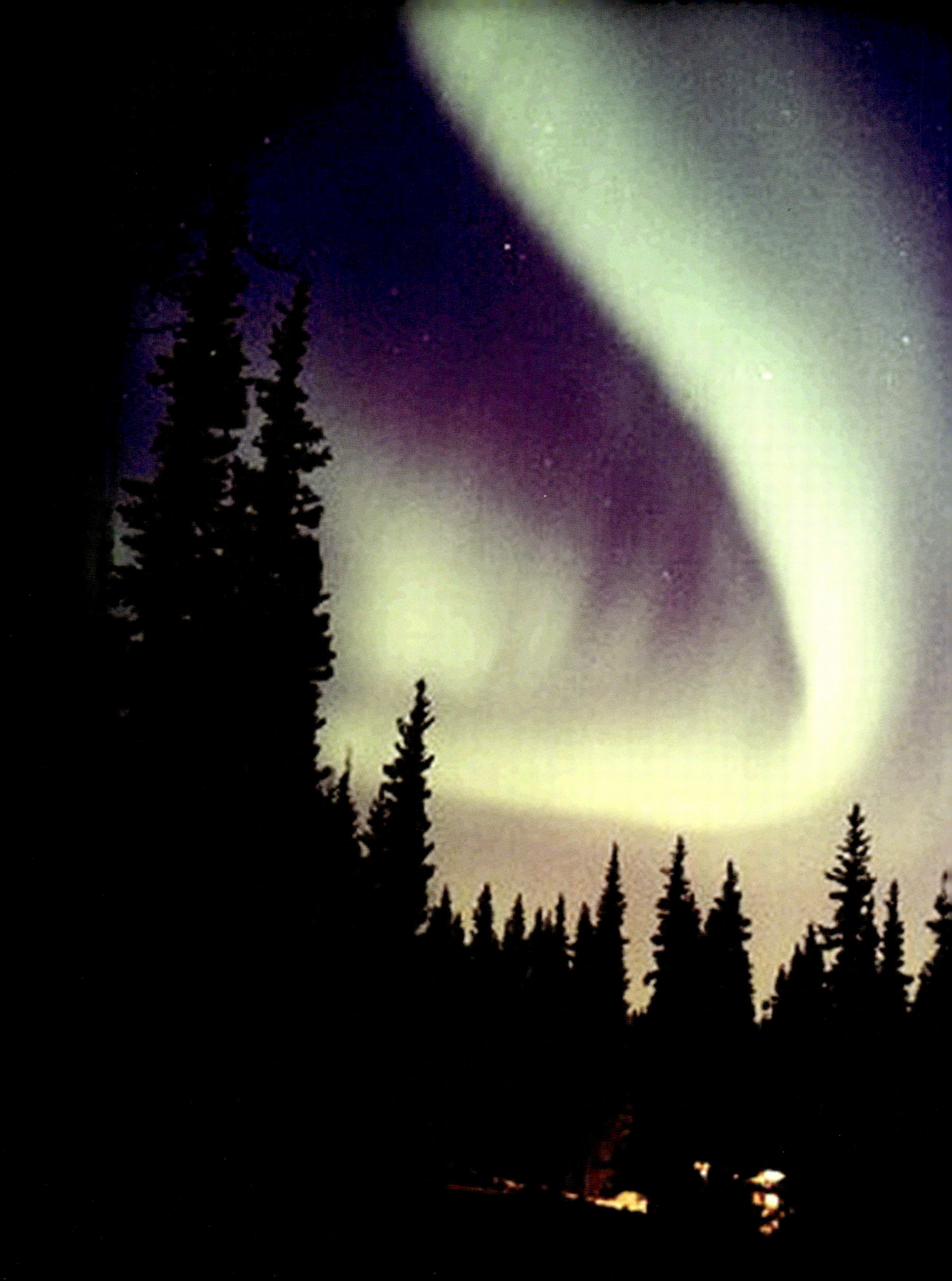

IMPACTS OF SOLAR VARIABILITY

Solar Causes, Terrestrial Impacts, and Societal Effects

Activity-related events on the Sun lead to societal consequences on the Earth through a causal chain of **solar events**, their physical **impacts** on the Earth, and the resultant **effects** of these impacts on human life and endeavor.

The initiating **causes** can be dynamic events at the Sun such as the expulsion of a CME or the eruption of a flare, or through more gradual forcing factors such as the Sun's 27-day rotation and the year-by-year variation in the extent of magnetic activity on the surface of the star. All of them impose variations—which can be sudden and large—in the amount of energy the Sun releases into space: in the intensity of electromagnetic radiation, the number and energies of solar atomic particles, or most often, in both of these.

Among the solar-induced **impacts** on the Earth—or in the now heavily-trafficked zone of space around it—are heightened levels of hazard in the near-Earth and space environment; disruptions and storms in the Earth's magnetosphere; the generation of strong electric currents that thread the atmosphere from top to bottom; changes in the ionosphere's reflective properties and in the chemistry of the stratosphere; and increased heating of the lower atmosphere, the surface of the Earth and the oceans.

The **effects** we feel on Earth or in near-Earth space cover a broad scope of national, commercial and personal activities and concerns. These range from satellite telephone calls and the nightly news on television to spaceflight, the transfer of electric power, national security, the transfer of financial and commercial information, weather and climate, and human health and well-being.

Some of the **impacts** of solar variability on different regions of the Earth and near-Earth space are summarized, with the solar **cause** of each of them, in the accompanying table. They are then explained, individually, in the text that follows it. A subsequent chart at the end of this section identifies some of the **human activities** affected by each of these **impacts**. The specific societal **effects** in each of these areas of human activity are covered, one by one, in the next section.

IMPACTS OF SOLAR VARIABILITY ON THE EARTH AND NEAR-EARTH SPACE AND THEIR CAUSES ON THE SUN

LOCATION	IMPACT
Near-Earth Space	**Greater flux of high-energy atomic particles** *CAUSE: eruptions of solar flares and CME-driven shock waves and the rise in their occurrence as solar activity increases* **Increase in highly-energetic EUV and x-ray radiation** *CAUSE: eruptions of solar flares and an increase in the level of solar activity* **An 11-year cycle in the Earth's receipt of galactic cosmic rays** *CAUSE: effect of the 11-year variation in solar activity and its impact on GCRs*
Magnetosphere	**Initiation of geomagnetic storms and sub-storms** *CAUSE: transfer of energy from CME-driven shock waves; high-speed streams of solar wind plasma and impulsive events on the Sun*
Upper Atmosphere	**Displays of the aurora** *CAUSE: entry of high-energy particles from the Sun or from the magnetosphere into the upper atmosphere, where they strike atoms of oxygen and nitrogen, resulting in the emission of visible radiation in discrete colors* **Changes in the temperature and dynamics of the thermosphere** *CAUSE: day-to-day and impulsive variations in the amount of solar x-ray and ultraviolet radiation that is absorbed at this level by neutral atoms of air* **Variability in the electrical properties of the atmosphere** *CAUSE: the direct transfer of energy from the solar wind into the magnetosphere, combined with impacts of high energy particles from solar flares and CMEs. These generate and sustain electric currents that couple the magnetosphere, thermosphere, mesosphere and stratosphere together, with impacts on the surface electric and magnetic field* **Fluctuations in the concentration of free electrons in the ionosphere** *CAUSE: changes in the Sun's x-ray and extreme ultraviolet radiation that is absorbed by neutral atoms of air in the thermosphere and ionosphere to produce free electrons and charged ions in the ionosphere*
Lower Atmosphere, Oceans, and Land Surface	**Changes in the abundance of ozone and other traces gases in the stratosphere** *CAUSE: fluctuations in solar ultraviolet radiation and high energy atomic particles from solar flares* **Changes in the temperature of the stratosphere** *CAUSE: solar-driven fluctuations in ozone that affect the amount of solar near-ultraviolet radiation that is absorbed at this level* **Changes in the temperature and other meteorological conditions in the troposphere** *CAUSE: persistent changes in the total and spectral radiation received from the Sun, including the near-ultraviolet, which by altering the amount of ozone in the stratosphere can affect the transfer of radiation throughout the lower atmosphere* **Changes in the surface and sub-surface temperature of the oceans** *CAUSE: variability in total and spectral solar radiation, which is absorbed within the topmost (photic) zone of the ocean, to a depth of about 650 feet in the clearest water* **Changes in the surface temperature of the solid Earth** *CAUSE: variability in the total and spectral radiation received from the Sun*

Impacts on Near-Earth Space

The inner heliosphere in which our shielded planet is immersed is an open range of hostile fire: shot through by both harmful short-wave radiation and potentially lethal atomic particles from the Sun and the cosmos.

The dosages of solar electromagnetic radiation and of energetic particles tend to track each other and are highly variable: both from year to year in the course of the 11-year cycle of solar activity, and from day-to-day and even minute-to-minute in response to solar rotation and impulsive events on the Sun and disturbances in the solar wind.

The amount of change in the intensity of emitted radiation or energetic particles is determined primarily by magnetic conditions on the surface and in the solar atmosphere. Since the temperature of the solar atmosphere increases with height, so does the energy and variability of particles and radiation that come from each level. For electromagnetic radiation this means that as we move upward from the photosphere through the chromosphere and into the corona the nature of radiation the Sun emits systematically shifts toward shorter, more energetic and more potentially damaging wavelengths.

The least variable parts of the solar spectrum—which change by no more than a few tenths of a percent from day-to-day or in the course of the 11-year solar cycle—are the visible and infrared portions: the light and heat that we receive from the Sun, which come from the relatively-cool and stable 10,000° F photosphere.

In contrast, the near-ultraviolet radiation emitted from the higher and more dynamic 20,000° chromosphere changes by up to several percent in the course of the solar cycle: a hundred times more variable than radiation in the visible and infrared. Extreme ultraviolet (EUV) and still-shorter wavelength x-ray radiation come from even higher, hotter, and more variable levels: from the transition zone and corona, where temperatures range from about 200,000 to about 2,000,000° F, and from magnetically-heated material in solar flares.

During the rapid three-to-four year rise in solar activity that follows each minimum in the 11-year solar cycle the radiation from the Sun in the highly-energetic EUV increases by 10 to 100 percent or more. In the even more energetic and potentially lethal x-ray region, the Sun's output can increase by factors of more than 1000.

Thus the most variable and impulsive regions in the spectrum of solar radiation are also the most energetic and potentially damaging. And although few particles and no short-wave radiation reach the surface of the Earth—not even at the summit of Mt. Everest—they are present in full strength in near-Earth space, just above the top of our thin blue atmosphere, where spacecraft fly.

The most energetic solar particles—those with energies per particle of a few hundred thousand to a billion electron volts or more—also come from explosive and eruptive events on the Sun. The flux of these, too, though highly variable in both energy and number, follows the ups and downs of the solar cycle with many more of the most energetic when the level of solar activity is high.

The flux of cosmic rays—with energies per particle of 10^8 to 10^9 or more electron volts—is also modulated by solar activity, though in the opposite way. When solar activity is high and conditions in the solar wind more turbulent, fewer cosmic rays reach the Earth. When solar activity is reduced, more cosmic rays arrive. Those with energies in the awesome range of 10^9 to 10^{20} electron volts, which are not uncommon, pass right through the heliosphere and our own upper atmosphere with the greatest of ease, seemingly oblivious to the presence of the Sun or how active it might be.

SOLAR ENERGY RECEIVED AT THE EARTH IN THE FORM OF ATOMIC PARTICLES

SOURCE	FRACTION OF ALL ENERGY RECEIVED FROM THE SUN	ENERGY INCIDENT PER ACRE AT THE MAGNETOPAUSE: WHICH TOTALS 5.5 MILLION WATTS	RANGE OF DAY-TO-DAY AND 11-YEAR CYCLIC VARIABILITY	COMPONENT THAT REACHES THE EARTH'S SURFACE
Coronal Mass Ejections	0.0018 % over periods of tens of minutes	Up to 100 watts	Occur sporadically: energy deposited varies by order of magnitude	None
High Energy Protons From Solar Flares	0.00015 % over periods of minutes to hours	Up to 8 watts	Occur sporadically: energy deposited varies by order of magnitude	Secondary particles of lower energy only
Solar Wind	.00002 %	1.2 watts	An order of magnitude	None
Galactic Cosmic Rays	.0000006 %	0.03 watts	20-25% 11-yr variation	None

This 11-year heliospheric modulation of cosmic rays, combined with the occurrence of explosive solar flares and CMEs that march to the same drummer, continually alter the character and level of risk in near-Earth space.

Magnetic Storms

A major impact of the never-ending flow of solar wind plasma is a reshaping and extension of the lines of force of the Earth's extended magnetic field, which, were the wind turned off, would revert to a simpler and more symmetric form.

The solar wind shapes the magnetosphere on its Sun-facing side by flattening the magnetopause and pushing it much closer to the planet it protects. On the nightside, the transfer of momentum from the solar wind extends the magnetosphere into a long tail that extends more than 20 times its undisturbed length.

These changes are more than superficial, for they can weaken the magnetosphere's defenses by making it easier for solar particles to intrude.

The principal *driver* of major disturbances in the magnetic field of the Earth—called magnetic storms—is the extended magnetic field of the Sun, also called the interplanetary magnetic field or IMF. Magnetic reconnection is also involved, which directly links the considerable kinetic energy of a fast or CME-accelerated solar wind stream to the magnetosphere and ionosphere. The overpowering pull of the solar wind also sets up patterns of large-scale circulation in the magnetospheric and ionospheric plasma which in turn provide the energy for geomagnetic storms.

The principal *impacts* on the Earth of magnetic storms are induced electric currents that run all the way from the magnetosphere to the Earth's surface, and even beneath the ground and on the ocean floor.

☉ ☉ ☉

Magnetic storms are often initiated by the arrival of a plasma cloud that has been accelerated by a fast-moving CME, or by high-speed streams in the solar wind. To initiate a storm either of these must bring with it solar magnetic fields with a significant south-pointing component. Both are often preceded by a developing shock wave that sets the storm in motion.

The effect at ground level of these sudden disturbances is an abrupt and sustained jump—sensed around the world—in the strength of the horizontal component of the Earth's magnetic field. These appear very clearly in continuously running records at magnetic observatories, or in erratic behavior of magnetic compass needles.

These tell-tale warnings—which have been recognized as such since the mid-1800s—signal the onset, far above, of a magnetic storm that will trigger a chain of disruptive physical, chemical and electromagnetic processes in the upper atmosphere, as well as major impacts on a continually increasing number of human activities. They can also flip the switch that turns on the northern and southern lights, in colorful and moving displays of the aurora borealis and australis.

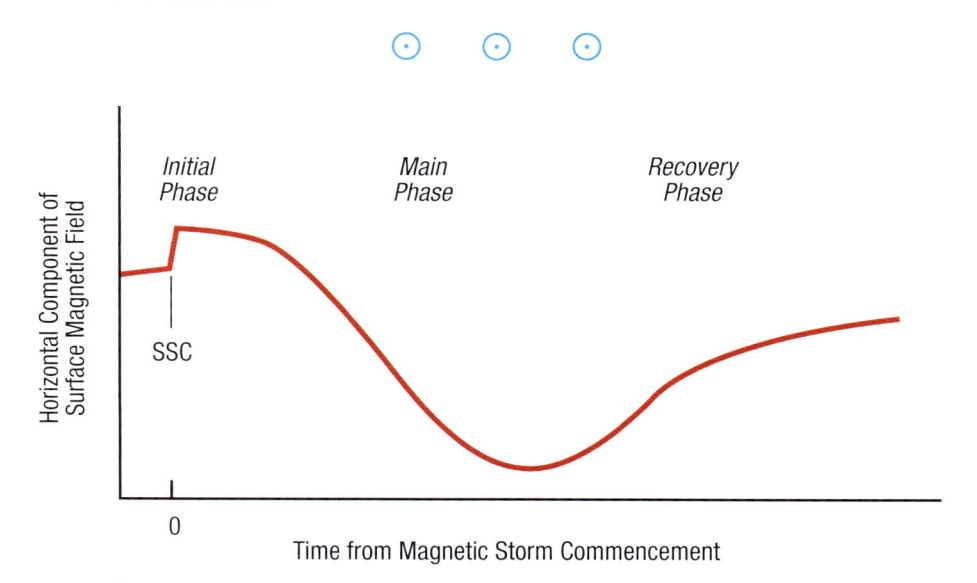

☉ ☉ ☉

[Figure: Graph showing horizontal component of surface magnetic field vs. time from magnetic storm commencement, with labeled Initial Phase (SSC), Main Phase, and Recovery Phase.]

Sequence of transient changes in the relative strength of the horizontal component of the Earth's surface magnetic field that together define a typical magnetic storm. In the initial phase, beginning with a sudden storm commencement (SSC), the field abruptly increases in strength by a distinct rise of a percent or so. This is followed by a slow and continuous weakening (main phase), lasting from a few to several hours that depresses the strength of the field by many times this amount, and a subsequent protracted rise (recovery phase) to its previous value, lasting hours to tens of hours.

Geomagnetic storms unfold in a well-established sequence of three steps, the first of which, described above, is the warning phase, which is also known as a sudden storm commencement or *initial phase*. At this time the lines of force of the Earth's field on the Sun-facing side are squeezed inward toward the solid surface of the planet, producing the observed sudden jump in the strength of the field. For several hours following this abrupt initiation, the Earth's magnetic field remains stronger than normal.

☉ ☉ ☉

During the *second* and so-called *main phase* of the storm, lasting hours to a day, the southward magnetic field in the cloud of solar particles makes contact with the Earth's oppositely-directed northward-pointing magnetic field. The two fields join in a process known as magnetic merging or reconnection and for a time the Earth becomes magnetically connected to an active region on the Sun, 93 million miles away.

The process creates open magnetic field lines rooted at either of the two magnetic poles of the Earth, some of which are then swept along the boundary of the magnetotail. With increasing distance from the Earth the magnetotail narrows, squeezing field lines of one magnetic polarity, rooted in the north magnetic pole, closer to those of the opposite polarity, attached to the south magnetic pole. At several fixed places, known as neutral lines, they are squeezed close enough to come in contact and join together, creating a long, closed magnetic loop with its two feet rooted back at the Earth in the two magnetic poles.

Streaming solar wind plasma that comes upon a neutral line in the magnetotail can pass through the magnetopause and be captured and accelerated in the magnetosphere, following one or other leg of the newly-formed closed magnetic loop back toward the Earth into a plasma reservoir, called the plasma sheet, in the magnetotail. There it mixes in with ions and electrons that have been carried up from the ionosphere.

⊙ ⊙ ⊙

Magnetic storms distend and disrupt the magnetotail, strengthening electric fields in the magnetosphere and ionosphere and accelerating particles in the magnetotail. Electric currents can at these times be shunted down into the northern and southern polar regions, where current-carrying particles collide with atmospheric atoms and molecules to produce displays of the aurora.

The systematic motions of positively-charged ions also intensifies an internal current in the magnetosphere known as the ring current that encircles the planet in a closed loop, confined to the plane of the magnetic equator, about 26,000 miles above the surface of the Earth.

At this altitude the ring current runs its circular course a few thousand miles beyond the farthest edge of the outer radiation belt, and about 8000 miles beyond the geosynchronous orbits where many modern spacecraft fly. One of its impacts on the magnetosphere is to suppress the strength of the Earth's magnetic field which, as we have noted, drops during the main phase of the storm to levels far below preceding conditions. The formation of a strong ring current is another defining feature of magnetic storms.

The powerful electric currents carrying more than a million amperes (!) in the high-latitude auroral region and the equatorial magnetosphere persist from a few hours to as much as a day, affecting not only the magnetosphere but the thermosphere and ionosphere as well. In the magnetosphere, the radiation environment for spacecraft is made more hazardous. In terms of societal effects, the changes invoked by magnetic storms perturb broad segments of modern activities. Included are civil and defense communications; commerce and industry; the operation and control of manned and unmanned spacecraft; the utility of geographic positioning systems and other aids to navigation, electric power systems, undersea cables and the operation of satellite telephones.

⊙ ⊙ ⊙

In the final, *recovery phase* of the storm, which lasts several days, the strength of the Earth's field—and with it, affected parts of technologically-aided life on the planet—gradually return to normal.

THE THREE STAGES OF A GEOMAGNETIC STORM

STORM PHASE	DURATION	CHARACTERISTICS
Initial phase (sudden commencement)	Several hours	Compression of the magnetosphere; abrupt onset and sustained high field strength
Main phase	Hours to a day	Gradual drop in magnetic field strength at low latitudes; displays of aurorae at high latitudes; systematic motion of charged particles within the magnetosphere; creation of an electrical ring current around the Earth; ionospheric currents at mid and high latitudes; effects on modern technology and society
Recovery phase	Several days	Slow return to normal field strength

The Aurora

The aurora borealis, and its counterpart in the Southern Hemisphere, the aurora australis, are celestial displays of glowing light that illuminate a considerable portion of the night sky: sometimes as rays but as often in the form of dancing ribbons that look like swaying curtains of glowing color. They typically persist for several hours if not all night and can appear in pure shades of green, red, blue, or yellow, or when mixed together, white. Though wispy and ethereal, aurorae can be easily seen with the naked eye with some displays a thousand times more luminous than the Milky Way, and as bright as the full Moon.

⊙ ⊙ ⊙

It can be no surprise that auroral displays were so often cited in legend, lore and myth. Nothing else that happens in the dark night sky is as colorful and spectacular, as large and so much in motion, as potentially mysterious or quite as awe inspiring. In northern parts of Iceland, Norway, Sweden, Denmark and Finland—where they appear out of nowhere almost every night—aurorae were seen, for millennia, as an undeniable demonstration of the close presence and power of supernatural forces. Moreover, their mystical origin seemed substantiated by the perception of accompanying sound: a soft and eerie crackling or rustling noise that many today still claim to hear when aurorae flutter like wind-blown curtains across the darkened sky.

In parts of Europe, Eurasia and the New World where they were more rarely seen, the appearance of an aurora was long perceived—as were comets and eclipses of the Sun or Moon—as a portent of awful things to come: or depending on the inclination of the observer, as a divine blessing of a just-completed happening, such as a birth or coronation or a battle fought and won.

A "coronal aurora," in which the locations of the light-emitting atoms of air define the general direction in space from which the incoming particles responsible for the induced emission appear to have come, making it look as though the aurora itself streams downward toward the Earth from a point-source in the sky.

As told a few years ago by the beloved historian Shelby Foote, such was the perception of many Confederate soldiers when on the cold dark night of December 13, 1862 they looked up to see what seemed to be a heavenly endorsement—proclaimed in shimmering banners of multi-colored light—of their bloody, hard-fought victory that day in the Battle of Fredericksburg, Virginia. For many of them, born and raised in the deep South, this was very likely the first aurora they had ever seen, and for some the last, as well.

⊙ ⊙ ⊙

The number of aurorae you might expect to see in a year, or a lifetime, depends first of all on where you are. Those who live within or close to the Arctic Circle—in northern Alaska, Canada's Northwest Territories, Greenland, Iceland, or in the northern parts of Norway, Sweden, Finland and Russia can watch ever-changing displays of the northern lights on almost every clear night. They occur less and less often the farther south one goes from there, such that within much of the continental United States the appearance of a bright aurora is a relatively rare phenomenon. And particularly so in the electrically-lit skies of the modern world.

In non-polar regions of the globe the number of aurorae that are there to be seen depends as well on the phase of the 11-year solar cycle: for it is CMEs and high-speed solar wind streams that initiate the chain of events, including magnetic storms, that turn most bright aurorae on and off. Nights on which bright aurorae were seen have been recorded in dynastic histories in China and Korea and in European diaries, personal and professional, since long before the advent of the telescope in 1609 and the discovery of the sunspot cycle in 1843. In succeeding years, reports of bright aurorae have tracked the 11-year solar cycle so well that the more ancient auroral accounts are now used to help extend what is known of the behavior of the Sun through the last 2000 years.

When averaged over the sunspot cycle, and in the absence of clouds, bright outdoor lights or a full Moon, about 100 auroral displays can be seen during an average year in Fairbanks, Alaska; 10 in Minneapolis; 5 in Chicago; 1 in Atlanta; and 1 per decade in Miami.

As the magnetic poles wander, as they always have, these zones of auroral occurrence will follow along, mile for mile. At the present time, the north magnetic pole lies some 800 miles south of the Earth's North rotational pole. Were it to continue its southward drift, the number of auroral displays seen in the sampled cities cited above would of course increase. Were it to reverse direction and drift north as far as the rotational pole, Chicago would see, on average, but one auroral display per year, Atlanta one per decade, and Miami perhaps one in a lifetime.

⊙ ⊙ ⊙

Auroral displays, which are also seen over the poles of Jupiter and certain other planets, are triggered by the arrival of streams of high-speed plasma from the Sun carrying southward magnetic fields. Auroral substorms, which are linked to the brightest and most dynamic aurorae, can occur during magnetic storms or quite on their own.

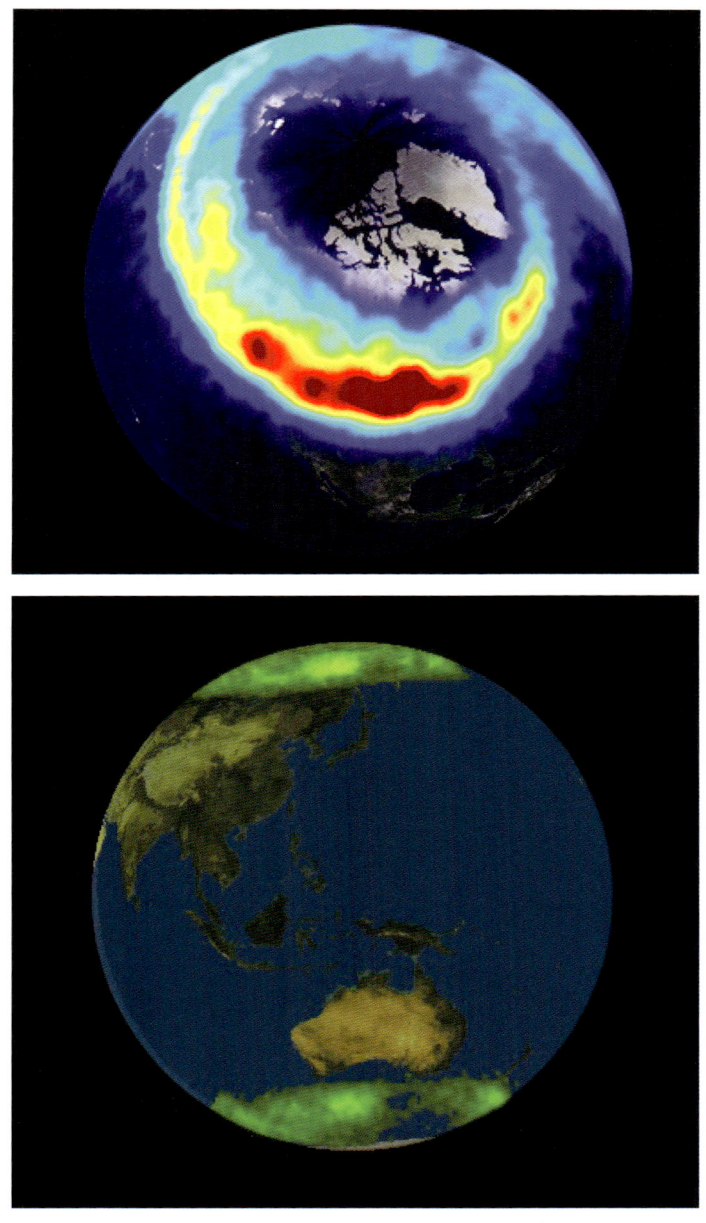

Images of the aurora borealis and australis, made from the POLAR spacecraft in a far-Earth orbit that crosses the poles of the Earth, offering distant views of all aspects of the entire spherical Earth.

Upper: the Northern Lights in full display, showing the complete auroral oval, here maximizing in brightness over northern North America, with portions of Greenland and northernmost Canada in sight within a more northern interior polar region where far fewer aurora occur.

Lower: the simultaneous display of the northern and southern lights, each centered on one of the two magnetic poles. Representations of the continents in these views of the night-side Earth are added artifacts.

Auroral substorms require an unstable stretched magnetotail with strong crosstail currents. Magnetic storms provide these conditions in spades and drive a complicated array of auroral activity. But much weaker and shorter duration southward fields from the Sun can trigger auroral substorms without the need for magnetic storm conditions to develop. In either case, a portion of the current across the magnetotail is suddenly shunted through the auroral ionosphere.

In the course of these sudden events some of the stored electrons are released from the magnetosphere's mighty grasp and propelled downward at high velocities into the thermosphere, ionosphere, and upper mesosphere. There, in the thin air at heights of 50 to 200 miles above the surface of they Earth they collide with neutral atoms and molecules of nitrogen and oxygen.

These collisions release photons of light, in the pure colors that are emitted from individual atoms that have been energized in the encounter. The brightest and most common colors in auroral displays are green and red, both of which come from oxygen atoms. Blue, next brightest, is produced by excited molecules of nitrogen.

⊙ ⊙ ⊙

And the mysterious whispering sounds, that much like UFOs, have long eluded proof of presence? The great distance (in most cases more than 100 miles) which separates the listener from the source, and the long delay (of up to ten minutes) that separates the initial arrival of light from the later arrival of sound makes any physical explanation highly improbable. The most likely answer, offered by the American physicist Elias Loomis in 1866, is that the perceived auroral sounds originate not up in the sky but here on the ground, between our own two ears.

We often hear what we expect to hear, and past experience—while witnessing fireworks displays or explosions, great or small—conditions our mind to *expect* that sound will accompany bright flashes of light: much as people nearly 2000 years ago, according to the Roman historian Tacitus, expected to hear a hissing sound while watching the setting Sun sink slowly into the ocean, just over the horizon, ten or twenty miles away; and often claimed they did.

⊙ ⊙ ⊙

Regular observations of aurorae from the vantage point of space allow a global view, seen from afar, of where they occur, confirming and extending what had been learned from the ground regarding the extent and characteristic patterns and differences of auroral displays.

Most aurorae in the Northern Hemisphere occur within an almost circular belt of latitude, 200 to 800 miles in width, known as the auroral oval, that is approximately centered on the north magnetic pole. A similar auroral oval, centered on the south magnetic pole defines the zone in which most aurorae occur in the Southern Hemisphere. The size of the auroral ovals is not fixed but elastic, their dimensions varying in response to the level of magnetospheric disturbances on any given day or year.

The north magnetic pole, as noted earlier, is now almost 800 miles south of the geographic pole and about 300 miles west of Greenland. At times of reduced solar activity, as at minima of the 11-year solar cycle, the northern auroral oval—which is centered roughly on that magnetic pole—can shrink in size until it is less than 1800 miles across: about the distance between Chicago and San Francisco. At these quiet times all of Alaska, most of Canada's Northwest Territories and all but the northern tip of Hudson Bay find themselves, like Denver or Havana, outside the shrunken auroral oval.

When magnetospheric activity increases, the northern auroral oval expands to a much larger size, often extending as far south as Seattle, Milwaukee, and Montreal, and—as was the case in the winter of 1862 (at the peak of the solar cycle that began at the minimum of 1856 and lasted through 1867)—on down to Fredericksburg, Virginia, where a war was being fought. During unusually energetic solar eruptions that trigger major magnetic storms the northern oval can stretch as far south as Mexico City and Panama.

☉   ☉   ☉

Another class of auroral activity occurs in association with the high-speed solar wind that blows outward from coronal holes. The wind from coronal holes is characterized by rapid variations in the imbedded magnetic field direction, with fields turning rapidly southward then northward then southward again over and over throughout the ten days it takes for a typical high-speed stream to reach the Earth. These short intervals of relatively weak southward magnetic field are sufficient to trigger substorm after substorm, lighting up the auroral oval repeatedly for a week or more. Strong high-speed streams reach a maximum in the declining phase of the solar activity cycle. Because of this, more energy pours into the auroral oval during years near solar minimum than near solar maximum when CMEs more seriously—but much less frequently—disturb the auroral oval.

The bright aurorae produced when the most energetic solar disturbances strike the magnetosphere occur almost entirely within the narrow band of the auroral oval. Seen from afar—as from the vantage point of space—the auroral oval appears as an expandable belt of brightness that slides up and down in latitude from day to day and year to year as magnetospheric activity wanes and waxes. Nor is the band equally wide all around the globe, for it is three or four times wider on the *night* side of the Earth than on the sunlit, *day* side where unseen aurorae also continually occur.

These weaker, daytime aurorae—most of which are caused by the precipitation of trapped particles—cannot be seen from the ground, against the glare of the daytime sky. But they are always there, and always have been: like the many flowers, in the words of Thomas Gray, that are born to blush unseen, and waste their sweetness on the desert air.

A nighttime view from the Space Shuttle *Discovery* while flying over the southern hemisphere. Visible, at the right, a reddish haze above white spikes of light at the curved southern horizon is a display of the aurora Australis, or southern lights: the counterpart of the northern lights seen near the opposite pole of the Earth. Both are features of the low thermosphere at heights of about 50 to 80 miles above the surface of the Earth, well below the Space Shuttle's altitude of about 200 miles. More brightly glowing at the left are ionized exhaust gases from the Shuttle's engines, and their light reflected off the spacecraft's vertical stabilizer.

Displays of another weaker class of aurorae—produced when high-speed solar wind electrons work their way into the magnetosphere through openings in the magnetotail—are an almost nightly phenomenon in polar and sub-polar regions. These polar cap aurorae occur poleward of the bounded region of the auroral oval, in the space that separates it from the magnetic pole, and are seen throughout the solar cycle.

TYPES OF AURORAL DISPLAYS

TYPE	INTENSITY	WHERE SEEN
Active aurorae associated with geomagnetic substorms	Brightest and most dynamic	Within the auroral oval
Daytime aurorae	Weak	At lowest latitudes in the auroral oval
Polar cap aurorae	Weak	Polar cap regions

Impacts on the Upper Atmosphere

Atoms and molecules of oxygen and nitrogen in the upper atmosphere—at altitudes from about fifty to several hundred miles above the surface—absorb almost all of the short wavelength, extreme ultraviolet and x-ray radiation that the Sun delivers to the Earth. In this transfer of energy, the Sun heats the air in the high atmosphere by a thousand degrees or more to create and sustain the thermosphere, 200 to 500 miles above the surface, while stripping bound electrons from neutral atoms and molecules to create the ionized layers within it which are called the ionosphere.

Since they are created and sustained by short-wave solar radiation, both the thermosphere and ionosphere are directly and instantly affected by any variation in the radiative energy that the Sun emits in this highly-variable region of the solar spectrum. Solar radiation in the extreme-ultraviolet varies in the course of the solar cycle or in response to solar eruptions by 10 to 20 percent, and in the x-ray region by orders of magnitude.

As a result, the thermosphere and ionosphere take the hardest hammering from changing solar radiation of any region in the atmosphere. They are also far and away the most disrupted and variable, including during the daylit hours when they must respond to the Sun's daily movement across the sky.

Changes in the amount of solar short-wave radiation provoke rapid responses in the temperature of the thermosphere. One of the largest perturbations is due to the daily rising and setting of the Sun, which in a matter of minutes drive the thermospheric temperature up or down through a range of from several hundred to several thousand degrees Fahrenheit. The pace of this diurnal switching—as different parts of the thermosphere are carried by the Earth's rotation into and then out of the sunlight—is not to be compared to the gradual changes in temperature that we sense at dawn or dusk on the surface of the Earth. In the thin air of the thermosphere the swings in temperature are far wider and are accomplished much faster.

TYPICAL DAY AND NIGHTTIME SUMMER TEMPERATURES IN SAN ANTONIO AND IN THE UPPER THERMOSPHERE

CIRCUMSTANCE	IN SAN ANTONIO, TEXAS	IN THE UPPER THERMOSPHERE
Nighttime	75° F	600° F
Daytime, quiet Sun	95	900
Daytime, active Sun	95	2100
Daytime, during a solar flare	95	3100

The day-lit half of the thermosphere must also respond, equally fast and vehemently, to any significant change in the energy that the Sun delivers in the EUV. The occurrence of a solar flare can raise the temperature of the thermosphere by another 1000° F or more; as can the evolving background change in solar activity, from years of minima to times of maxima in the cycle.

Adding extra heat to the already hot thermosphere causes it to swell and expand upward. For example, the 1500° difference between nighttime and daytime temperature due to the Earth's rotation introduces a hemispheric bulge on the Sun-facing half of the thermosphere: a congenital lopsidedness which has been a feature of our planet from the time it had an atmosphere.

In addition to this daily oscillation, the vertical *extent* of the thermosphere has for billions of years been rising and falling on other scales of time: breathing out and breathing in—as solar activity waxed and waned—in response to day-to-day and year-to-year changes on a star 93 million miles away.

For the most part, these heat-driven waves at the top of the atmosphere were until the middle of the last century largely unnoticed or ignored—like the heaving of the middle ocean before the days of sail. But when man-made satellites began to circle the Earth, these variations came to be important.

When the thermosphere expands upward, it intrudes into the more tenuous regions through which many Earth-orbiting spacecraft fly. For the *Hubble Space Telescope*, the *International Space Station* or any of the other spacecraft orbiting the Earth in the 150 to 400 mile range, any increase in the density of the medium through which it flies will slow it down. This in turn reduces the diameter of its orbit and the height at which it flies, and hastens the day when it will plunge into the deeper atmosphere and return, in pieces, to the surface of the Earth.

Perturbing the Earth's Electric Field

Solar-driven changes in the temperature, density and dynamics of the thermosphere can also perturb the Earth's electric field: a fundamental property of the planet which we witness from time to time in the course of thunder and lightning storms.

Flashes of lightning mark times and places where the negatively-charged surface of the Earth is momentarily connected, through an electrical discharge, to the storm-generated, positively-charged upper layers of clouds that float above it. These spectacular happenings are but the most visible element of a global electric circuit that couples all of the atmosphere, from top to bottom, to the Earth beneath it: one of the few *direct* connections that link the top of the atmosphere to the biosphere, hundreds of miles below.

Thunderstorms are one of the three mechanisms that generate the electrical power that establishes and maintains a voltage of about 300,000 volts between the negatively-charged ground and the positively-charged ionosphere. One of the others is the dynamo that generates electric currents when the solar wind interacts with the Earth's magnetic field.

Winds in the thermosphere power a third electric current generator. These are driven, like the winds we feel at the surface, by differences from place to place in the local temperature and density, which in this case are induced by variations in the intensity of solar EUV radiation. At these lofty heights the thermospheric material that is carried in the winds is weakly-ionized plasma: ions and free electrons that act as an electrical conductor and current generator when they move through the lines of force of the Earth's magnetic field.

The currents added to the global circuit from any of these sources affect the difference in voltage between the ionosphere and the Earth's surface. Charged atomic particles, whether from the Sun or from more distant cosmic sources, can bring about a similar effect by reducing the electrical *resistance*—normally almost infinite—in the layers of air that separate the ionosphere from the solid or liquid surface of the planet. Alone, or acting together, these two impacts—one from the Sun's EUV radiation, the other from incoming charged particles—directly affect the global electric circuit, and perhaps the number of thunderstorms.

Restructuring the Ionosphere

The ionosphere, or ionized atmosphere, is the name given to the concentrated layer of charged particles—electrons and ionized atoms and molecules—that extends from the top of the mesosphere through the lower part of the thermosphere: from 40 to about 400 miles above the surface of the Earth, with a maximum charge density at a height of about 200 miles.

The electrons and ions are created and sustained there through the action of solar short-wave radiation, solar particles and cosmic rays on neutral atoms and molecules of nitrogen and oxygen, the most abundant species in the atmosphere. In the process of these encounters the neutral atoms and molecules are deprived of some of their bound electrons and become electrified, or ionized.

The outstanding societal effects of the Sun's creation and control of the ionosphere are its impacts on electronic communications of all kinds and at almost all frequencies, including radio, television, satellite telephone, navigation, data transfer systems, spacecraft control and countless military and other national security systems.

Disruptions in these and other telecommunications systems are caused by solar-driven changes in the concentrations of charged particles at different heights in the ionosphere, which alter how electromagnetic waves are reflected, absorbed or allowed to pass through it.

⊙ ⊙ ⊙

The number of electrons and ions in the ionosphere is determined by an ongoing give and take between their rate of production—governed by the varying intensity of solar short-wave radiation and incident particles—and the rate at which the newly-freed electrons and ions once again *recombine* to form reconstituted particles of neutral charge.

Thus, as a portion of the ionosphere is carried by the Earth's rotation into the darkened half of the planet, the rate at which electrons and new ions are produced falls rapidly toward zero: as does the number of charged particles within it, such that by the end of the night the ionosphere's structure and composition have changed dramatically. So much so that most of the lower ionosphere—beneath a height of about 70 miles—will have vanished, leaving only an upper part that extends high into the thermosphere, called the F-region.

Soon after sunrise, incoming sunlight restores the lost lower layers while increasing the thickness and density of charged particles in the much higher all-night region by about a factor of ten: in all, a dawn and dusk ritual, high in the sky, which has been acted out, unseen, since long before dinosaurs first appeared on the planet.

The reality of day-to-night changes in the structure of the ionosphere is quite apparent to anyone who when tuning in after dark to AM radio hears unexpected broadcasts from far away stations with strange sounding names: words, often in foreign tongues, that have been bounced and rerouted part-way around the Earth by the high nighttime ionosphere.

Other more disruptive and less predictable changes in the density of the ionosphere are the direct impacts of solar variability: specifically, changes in the amount of energy that the Sun releases in the form of EUV and x-ray radiation and the impacts of charged atomic particles. During a solar flare, for example, the ultraviolet and x-ray radiation from the Sun increases from 10 to more than 100%, as it does in the course of the 11-year solar activity cycle, and the energies of incoming solar particles by as much or more than that.

☉ ☉ ☉

Since the late 1920s, the structure of the ionosphere has been described in terms of three distinct horizontal *layers*, distinguished by composition and vertical extent and labeled, from the lowest upward, the D, E, and F regions.

All three are made up of ions and free electrons, but the expected lifetimes of any of these short-lived atomic particles vary from layer to layer, due principally to the ever-decreasing density with height in the neutral atmosphere. In the F-region, which extends from 100 to 400 miles above our heads, the air is so thin that ions and free electrons less frequently come in contract with other particles, and as a result can exist an hour or so before they recombine.

The greater concentration of particles in the E-region (60 to 80 miles high) shortens the expected lifetimes of ionized particles to a few minutes; and in the most dense and crowded D-region (40 to 60 miles high) charged atomic particles remain that way for but a few seconds. It is because of these short lifetimes that densities of particles in the D-region and a part of the E soon fall to zero at nightfall.

Two factors work to keep the F-region of the ionosphere from disappearing altogether in the dark of the night. The first, noted above, is the fact that ions, once created, survive a while longer in less dense regions of the atmosphere before recombining: an extended lease on life that lengthens with increasing height in the ionosphere.

The second is the replenishment of a fraction of the lost ions by the impact of nighttime cosmic rays on neutral atoms and molecules, supplemented by enfeebled solar ultraviolet radiation that has been scattered and redirected at high levels into the darkened half of the ionosphere from the adjoining daylit half. This weak leakage of light from day to night is not unlike the diffuse nighttime glow we commonly see, just over the horizon, from the lights of a neighboring city.

Some ions and free electrons are found in the upper atmosphere both above and below the portion that we call "ionosphere." With increasing altitude more and more of the neutral atoms and molecules become ionized, and the atmosphere is transformed from wholly neutral to almost completely charged.

The gradual changeover from neutral to ionized begins at an altitude of about 35 miles—not far above the stratopause—and continues all the way to the top: where the "atmosphere" ends and the exosphere and magnetosphere begin. At this ill-defined boundary electrons and ions flow both up and down: many of those carried up into the magnetosphere by the daytime thermal expansion of the thermosphere are drawn back down again at night as the thermosphere cools and shrinks.

What distinguishes the ionosphere from other regions of the atmosphere that also contain electrons and ions is the *number* of charged particles which are found there. For although the *ratio* of ionized to neutral particles is less in the ionosphere than in the high thermosphere and exosphere, there are many *more* of them per cubic inch or cubic mile. It is also in this unique band of altitudes—between about 40 and 400 miles above the surface—that the optimum is reached between the *intensity* of solar short-wave radiation (greater the higher you go) and the *number* of neutral particles to be ionized, which goes the other way.

⊙   ⊙   ⊙

But why D, E, and F? These familiar identifiers of the three layers in the ionosphere were assigned in the late 1920s, in the course of the first remote soundings of the region. In the heady days of initial discovery, it was thought prudent to hold the *first* three letters of the alphabet in reserve, in case additional, lower layers might be found. But as it turned out, neither these nor any of the letters that follow F were ever called into service.

THE LAYERED IONOSPHERE

REGION	HEIGHT IN MILES	CONTENT	TYPICAL LIFETIMES OF PARTICLES	PRINCIPAL IONS	MAIN SOURCE OF IONIZATION
F	100-400	electrons, ions	1 hour	O^+	solar EUV radiation
E	60-100	electrons, ions	1 minute	NO^+, O_2^+	solar EUV and x-ray radiation
D	40-60	ions and in the daytime, electrons	1 second	NO^+, O_2^+	solar EUV and x-ray radiation; solar particles and cosmic rays

Disturbing the Biosphere: The Lower Atmosphere, Oceans, and Land Surface

The most energetic and highly variable radiation from the Sun—in the short-wave EUV, x-ray and gamma–ray spectrum—is wholly absorbed in the upper reaches of the atmosphere and never makes it down into the biosphere, where all of life is found. Nor do any of the highest energy atomic particles that rain down on the Earth from the Sun and other cosmic sources.

What does reach the lower atmosphere, oceans, and solid surface of the planet are less energetic particles, light (visible radiation), heat (the infrared), and a small portion of the invisible near-ultraviolet: in all, about 60% of the energy the Sun delivered at the top of the atmosphere.

Some of the 60% is reflected back into space by clouds and land and sea and ice, but by far the largest part of this—about half of what the Sun delivers—is utilized in the biosphere. Among these tasks are heating the oceans and the solid surface, and through these, the lower atmosphere; creating winds and clouds and providing rain; regulating CO_2 and other greenhouse gases through the Earth's solar-driven carbon cycle that returns some of these pollutants to the oceans and ocean sediments, where through plate tectonics and volcanism they are eventually released; driving the hydrologic cycle that makes the rivers flow; fueling photosynthesis; establishing and sustaining the ozone layer in the low stratosphere; coloring the sky; and illuminating much of what we do.

Because the Sun's outputs of visible, infrared and near-ultraviolet radiation all vary to some degree in response to solar activity, all of these terrestrial services are potentially affected by changes on the Sun. In most of them, including all that involve visible or infrared radiation from the Sun, the immediate impacts of solar variability are far smaller than the changes induced by internal forcing in the system. Among these non-solar sources of variability are the day-to-night changes that result from the Earth's rotation; the perennial march of the seasons; recurrent ocean-atmosphere interactions such as El Niño/La Niña; the varying absorption of clouds, atmospheric aerosols and pollutants; and in the longer term, the introduction of greenhouse gases of human origin.

ORIGINS OF SOLAR RADIATION AND AFFECTED REGIONS OF THE EARTH'S ATMOSPHERE

SPECTRAL RANGE	SOLAR ORIGIN	AFFECTED PART OF THE EARTH'S ATMOSPHERE	WAVELENGTH RANGE (IN NANOMETERS OR MICRONS µ)
Visible and Infrared	Photosphere	Troposphere, stratosphere	400 nm to 10 µ
UV	Upper photosphere and chromosphere	Stratosphere, mesosphere, lower thermosphere	120 to 400 nm
EUV	Chromosphere, transition zone	Thermosphere; E and F layers of the ionosphere	10 to 120 nm
X-ray	Corona	D and E layers of the ionosphere	0 to 10 nm

Since solar radiation in the visible and infrared portions of the spectrum varies so little—less than half a percent in the course of a day, a year or the 11-year cycle—any of these competing weather and climate drivers can easily overwhelm and hide more subtle changes of solar origin, and particularly on shorter scales of time: minutes or months and even years.

As a result, it is often only in the longer term—through slower and more persistent forcing—that the deeper sounds of the Sun emerge through the noisy background chatter of shorter and more ephemeral perturbations.

More immediate and readily recognized are the impacts on the lower atmosphere of the near-ultraviolet solar radiation that reaches the lower stratosphere and upper troposphere. In the region between about ten and thirty miles above the surface, the absorption of solar ultraviolet radiation by molecular oxygen creates a small amount of ozone (about 1 part ozone to 10 million parts of air) which, though variable from place to place and time to time, is sufficient to shield life on the surface of the Earth from more lethal doses of solar ultraviolet rays.

Atmospheric ozone is also the creator of the sinuous vertical temperature profile of the Earth's atmosphere. This it does by absorbing solar short-wave radiation, slowing the steady drop in temperature with height in the troposphere, and at the tropopause, turning it completely around so that the temperature of the air *increases* with height through the next 20 miles of altitude. The result is the Earth's warm stratosphere: a feature of our planet that is unique in the solar system.

Both the thickness of the ozone layer and the temperature structure of the stratosphere are directly affected by changes in solar activity since they are each driven by solar radiation in the near-ultraviolet. But what the Sun giveth it also taketh away, for solar radiation in an adjacent region of the near-ultraviolet breaks ozone molecules apart.

Also at work is the destruction of ozone by chlorine, almost all of which finds its way into the stratosphere in the form of industrially-produced chlorofluorocarbons (CFCs), and by the catalytic influence of nitrogen oxides—much of which is produced by agriculture and the burning of fossil fuels.

The well-established solar control of stratospheric ozone offers as well a potential mechanism that might link solar variability to weather and climate, through connections that tie meteorological conditions in the stratosphere to those in the troposphere.

What is undeniable is the role of solar infrared radiation in establishing the temperature and circulation of the lower atmosphere, the temperature of the surface and subsurface ocean, and the temperature of the Earth's surface. Since the Sun's infrared radiation varies systematically in step with the solar cycle—rising slightly with increased activity and falling as it wanes—its influence should and does show up in records of all of these weather parameters.

IMPACTS OF SOLAR VARIABILITY ON THE EARTH AND NEAR-EARTH SPACE AND THE HUMAN ACTIVITIES THAT THEY ADVERSELY AFFECT

SOLAR IMPACTS \ HUMAN ACTIVITIES	AIR TRAVEL	HUMAN SPACE FLIGHT	OPERATION OF SPACECRAFT AND SPACE EQUIPMENT	OBSERVATIONS OF THE EARTH FROM SPACE	TELECOMMUNICATIONS AND NATIONAL SECURITY	GEOGRAPHIC POSITION FINDING AND NAVIGATION	ELECTRIC POWER TRANSMISSION	OPERATION OF OIL AND GAS PIPELINES	GEOLOGIC SURVEYS AND EXPLORATION	CLIMATE
INCREASED LEVEL OF UV AND X-RAY RADIATION	●	●	●	●	●	●				●
INCREASED FLUX OF ATOMIC PARTICLES	●	●	●	●	●	●				
MAGNETIC STORMS	●				●	●	●	●	●	
AURORAL DISPLAYS										
THERMOSPHERIC TEMPERATURE		●	●							
INDUCED ELECTRIC CURRENTS			●	●	●	●	●	●	●	
IONOSPHERIC CHANGES	●	●	●	●	●	●				
STRATOSPHERIC OZONE										●
• AIR TEMPERATURE										●
• OCEAN TEMPERATURE										●
• LAND TEMPERATURE										●

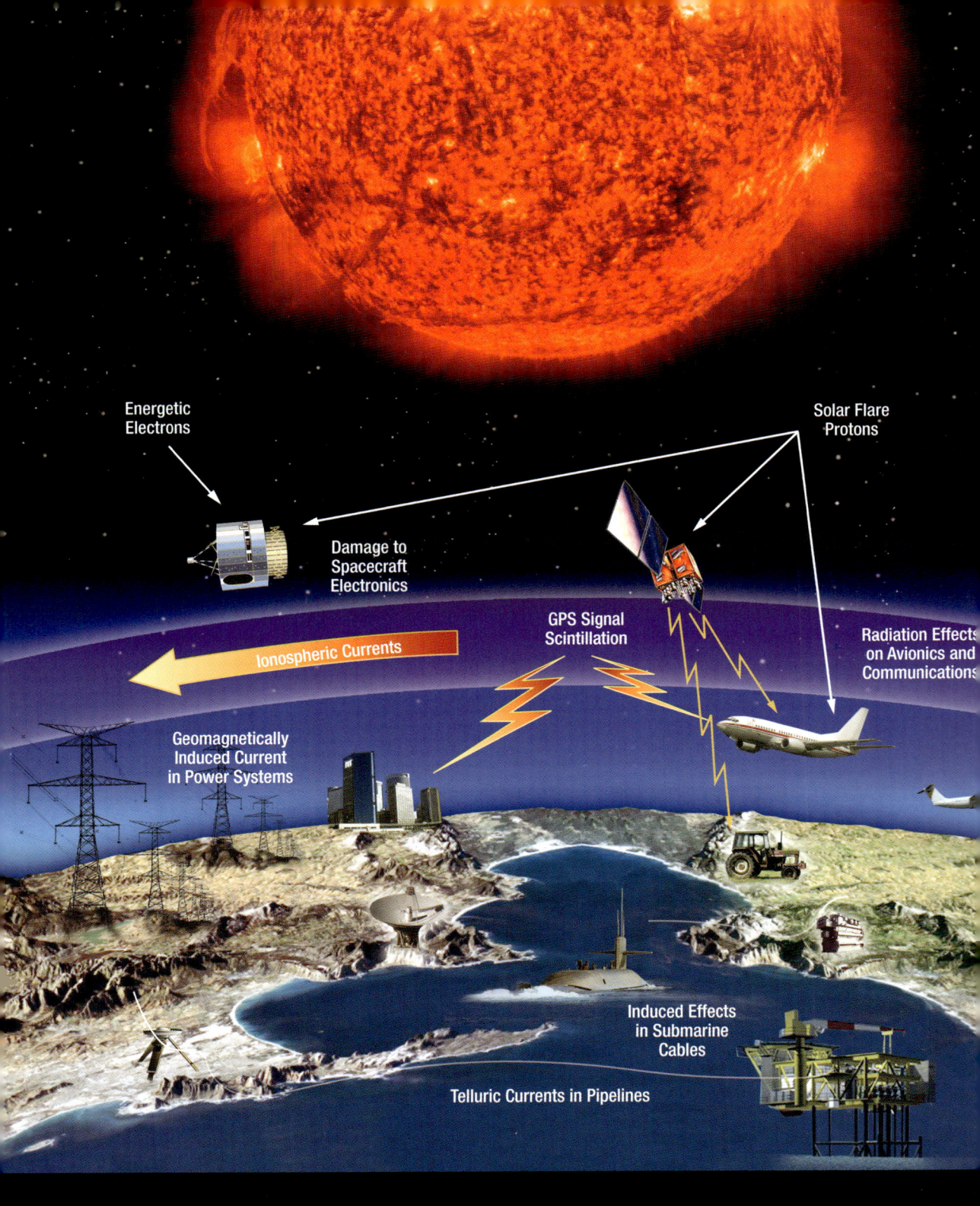

Some of the ways by which solar eruptions and the geomagnetic storms which they induce affect our lives and livelihood on th

EFFECTS ON HUMAN LIFE AND ENDEAVOR

What Is Affected

In what ways are we affected when solar disturbances alter conditions in the Earth's near-space environment, in the magnetosphere and ionosphere, the atmosphere, and the oceans and solid surface of the planet?

Two hundred years ago, when sunspots were all that was known of solar variability, it was suspected that their coming and going were in some ways affecting the weather and the number of auroral displays that were seen, athough neither connection had been verified. By 1950, however, when telegraphy, transoceanic cables, telephones, wireless radio, radar and more extensive electric power distribution systems had come upon the scene, the societal impacts of solar activity had reached far beyond possible ties to the weather into other areas of societal concern. Foremost among these were disabling effects on radio and other electronic communications, and disruptions of electric power distribution systems.

By the late 1950s, and particularly after our entry into space, the list of societal consequences rapidly increased in number and importance, as they have, almost exponentially, since then.

Today the heavy commercial and military reliance on spacecraft—particularly those in more hazardous high-Earth orbits—and the plans for a manned return to the Moon and for possible human exploration of distant Mars have vastly increased our vulnerability to the changing moods of the Sun. As has the switch in the last decade from custom-produced "space-hardened" electronic components to those that are available commercially, off-the-shelf.

In this section we summarize and then describe some of the specific ways through which the Sun's impacts on the Earth affect a broad range of human activities.

Some Specific Societal Effects

AIRCRAFT TRAVEL
Operational flight control and other aircraft communication outages;

exposure of passengers and crew on high altitude polar flights to mild dosages of CME-accelerated particles, high-energy particles from solar flares, and GCRs; and disruption or failure of essential ground-to-air and air-to-ground communications

HUMAN SPACE FLIGHT
Increased exposure of astronauts, both within a spacecraft or, with greater risk, while outside it, to high energy particles from solar flares and CMEs. During EVAs or when on the Sun-lit surface of the Moon or Mars, astronauts are also highly vulnerable to enhanced x-ray emission from the Sun

OPERATION OF SPACECRAFT AND SPACE EQUIPMENT
Electrical charging of the surface of the spacecraft resulting in degradation of the metallic surface; induced charging of cables and surfaces within the spacecraft, leading to deleterious impacts on the operation of computer memory and processors and to damage or failure of semiconductor devices in electronic components and other instruments; orbital perturbations and accelerated orbital decay due to the solar-driven expansion of the thermosphere; degradation of the surfaces of solar cells; disruption of spacecraft attitude-control; and interference with low-altitude satellite tracking

OBSERVATIONS OF THE EARTH FROM SPACE
Interruption and degradation of data used in preparing daily meteorological forecasts, in tracking and monitoring hurricanes and other storms, in forecasting expected crop yields, and for surveillance and other national security purposes

COMMUNICATIONS AND NATIONAL SECURITY
Ionospheric disturbance of radio transmissions of all kinds and in almost all radio-frequency bands, affecting nearly every form of electronic communication, including satellite telephones, network television, operation of communications satellites, transmissions to and from orbiting spacecraft, and essential military communications and radar systems. Other disturbances arise from induced electrical currents in long under-sea telecommunication cables, long-haul telecommunication lines, and certain fiber-optics systems

GEOGRAPHIC POSITION FINDING AND NAVIGATION
Errors and reduced accuracy in GPS systems affecting navigation and position finding at sea, in the air, and on the surface of the Earth; malfunction of other navigational aids; and the introduction of magnetic compass errors

ELECTRIC POWER TRANSMISSION
Power black-outs, brown-outs and disruption of electric power grids, triggered by solar-induced currents in power lines and overheating and eventual failure of

power-line transformers; curtailing the cost-saving ability of electric power grids to move cheaper electricity from available sources to users at more costly sites

OPERATION OF OIL AND GAS PIPELINES
Solar-induced electric currents in long pipelines at high latitudes, leading to corrosion and the ultimate failure of metal and welded joints

GEOLOGICAL SURVEYING AND PROSPECTING
errors and reduced accuracy in geological survey work, including prospecting for minerals and oil, due to solar-driven perturbations and the impacts of magnetic storms on the Earth's surface magnetic field; operational failures in drilling operations that rely on the Earth's magnetic field for directional reference

REGIONAL AND GLOBAL CLIMATE
Increase in the mean surface temperature of the Earth and the surface and subsurface temperature of the oceans, affecting atmospheric circulation and precipitation; exacerbation or amelioration of other agents of climatic change, including El Niños and human-induced greenhouse warming

Exposure of Aircraft Passengers and Crews

Solar activity can affect the safety of high altitude jet aircraft flight in two different ways. The first and most often encountered is interference in essential ground-to-air and air-to-ground communications. The second is the exposure of aircraft passengers and crews to potentially-harmful high-energy atomic particles.

Passengers and crews in high-flying aircraft are fully shielded from potentially-harmful solar ultraviolet and x-ray radiation by the metallic skin of the aircraft, just as astronauts are shaded from more intense sunlight at spacecraft altitudes.

This is not the case, however, for energetic particles from solar flares, CMEs and cosmic sources that have made their way into the upper troposphere and lower stratosphere, where aircraft fly. For these fast-moving and extremely energetic particles, the metallic skin and frame of an aircraft present almost no obstacle at all. Sufficiently energetic cosmic rays and solar particles will also pass through the aircraft interior as well as clothing and human skin like high-speed bullets through a paper target, to continue on at high velocity into human tissue and cells.

There they can ionize some of the atoms of which living cells are made, and—in sufficient dosage—alter the structure of impacted cells, and through these changes introduce mutations in the ensuing generations that these cells produce. Sufficient exposure to this kind of so-called ionizing radiation, (whether at one time or accumulated in the course of many months or years) can produce different forms of radiation sickness and possibly lead to cancer.

The degree to which crews or passengers are affected on any jet flight depends on the flight trajectory, including altitude and the regions of the Earth over which it flies. Air travel at conventional jet aircraft altitudes and within middle or low latitudes—like most flights within the lower 48 states—are but little affected.

☉ ☉ ☉

More at risk are flights that spend a significant amount of time over polar and sub-polar regions, where energetic solar particles more easily stream down into the atmosphere. Most of these incoming particles lose much of their original energy high in the atmosphere, through collisions with atoms and molecules of air, but some of the second generation particles that are produced in these collisions continue downward into the altitudes where jet aircraft fly.

The number of downward streaming secondary cosmic rays reaches a maximum at an altitude of about twelve miles or 66,000 ft: but a few miles above the highest altitude at which today's jet aircraft fly. Below that, depleted by thicker air and more frequent collisions among particles, the number begins to fall.

Thus, in the range of altitudes at which commercial, business, and military jet aircraft fly—between about 30,000 and 50,000 ft—the expected dose of secondary cosmic particles increases markedly with altitude. A jet aircraft climbing from 30,000 to 40,000 ft (the range of altitudes of most commercial airliners) triples its expected exposure to cosmic rays. Climbing above 40,000 to the rarified altitudes at which the next generation of commercial airliners will fly—between 50,000 and 60,000 ft—will increase the dosage by another factor of three or four.

A map of the world showing the expected dosage of energetic particles at a given jet aircraft altitude would portray an extensive zone of significant risk centered on each of the two magnetic poles of the Earth. Due to the considerable offset of the north magnetic pole (today about 1000 miles south and west of the North *rotational* pole) this zone of greatest exposure reaches farther down in latitude in western North America than elsewhere in the Northern Hemisphere.

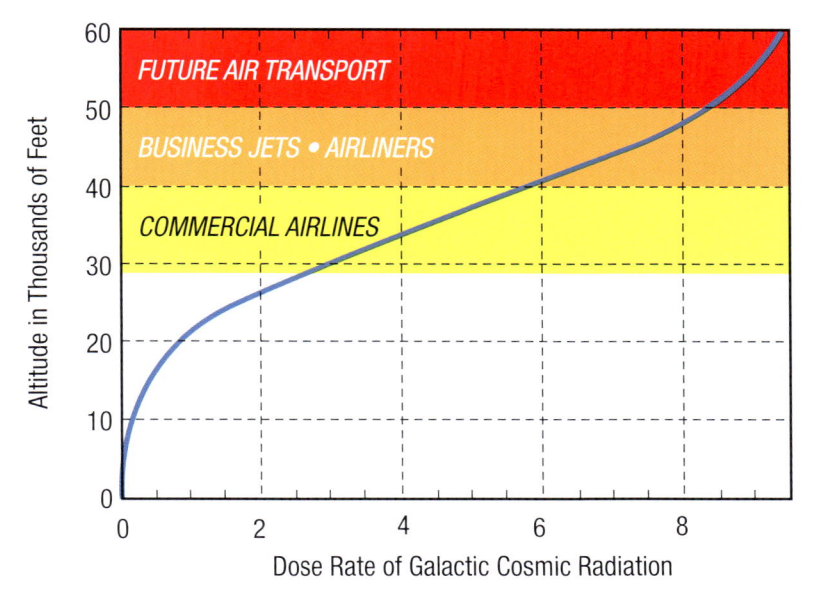

The dose of damaging radiation from galactic cosmic rays expected at different aircraft altitudes, from the surface of the Earth to 60,000 ft., showing the altitude range at which different types of aircraft now operate. A prevailing trend in future aircraft planning and design is to operate at altitudes in the now little-used band between 50,000 and 60,000 ft. in order to increase the range of ever larger aircraft and to utilize less congested air space. Any change in this direction will sharply increase the exposure of aircraft passengers and crews to high energy galactic particles. The curve shown here applies to low and middle latitudes; the dosage will increase as a function of higher geographic latitude, with highest dosages in polar and sub-polar regions.

This southward bulge of the more hazardous zone affects all who travel at high altitudes over Canada and as far south as the Great Lakes and the northern tier of states in our own country. Most exposed are passengers and crews on long-distance, intercontinental flights—such as those from New York to Tokyo, or Seattle to London or Frankfurt—that follow great circle routes that carry them over sub-polar regions in northern Canada or Greenland. In the course of five of these polar crossings a passenger or crew-member will have typically absorbed an amount of high energy particle radiation that exceeds the maximum recommended *yearly* dosage.

The routes followed by the *Concorde*, which took it back and forth across the Atlantic at *middle* rather than high latitudes, mitigated some of the added risk that comes with flying at an altitude of about 58,000 ft (compared to 30 to 50,000 feet for conventional commercial airliners.) The *Concorde* was also helped in this regard by its supersonic speed which—like running instead of walking through the rain—considerably shortened the exposure time.

☉   ☉   ☉

The radiation dosage encountered in most commercial jet-aircraft flights is no more than what one expects from a medical x-ray; although in a flight that happens to cross polar and sub-polar latitudes at the time when high-energy solar protons arrive, the dosage would considerably exceed that amount. In any case, in most single encounters of this kind, immediate risks to the health of passengers and crew are not severe.

These concerns increase with repetition, as in the case of passengers or crews who regularly or frequently fly on intercontinental flights that take them across these more exposed polar regions. Most at risk are pregnant passengers or crew members, aircraft crews in general, and the half million frequent fliers who log more than 75,000 miles—three times around the Earth at the equator—year after year.

The intensity of particle radiation at jet aircraft altitudes can be ten to fifteen times higher than normal after the eruption of major flares and CMEs, with highest dosages in those cases when streams of energetic particles from a major solar eruption happen to reach the aircraft while it is passing over polar and sub-polar regions. The most energetic of these particles are protons from major solar flares with single-particle energies in the range of a million to a billion electron volts. The more energetic of these can reach an aircraft in less than twenty minutes after the flare was first sighted, limiting the time available for aircraft to be diverted to reduce the impact of these particle incursions on air-to-ground and ground-to-air communications.

AIRCRAFT OPERATIONAL ALTITUDES IN RELATION TO ATMOSPHERIC FEATURES

ATMOSPHERIC FEATURE / TYPE OF AIRCRAFT	DISTANCE ABOVE THE EARTH'S SURFACE IN MILES	IN THOUSANDS OF FEET
Average height of tropopause	6	32
Commercial airliners	6 - 8	32-42
Business jets	8 - 10	42-53
Tropopause over equatorial latitudes	9 - 10	48-53
Supersonic aircraft (Concorde)	11	58
Planned future air transport	10 - 12	53-63
Maximum dose of secondary cosmic rays	12.5	66
Deepest penetration of primary GCRs	25	132

The possible effects on health of accumulated exposure are of sufficient concern that airlines and military flights routinely alter flight paths and altitudes to minimize the effects of particle radiation on communications; maintain records of each pilot's and crew member's career exposure; and rotate flight crews so as to limit their cumulative total. For these reasons, the European Union has adopted laws that mandate a certain amount of radiation monitoring on commercial airline flights.

Risks to Manned Space Flight

Three perils await space travelers who venture beyond the atmospheric and magnetospheric shields that for billions of years have allowed life to evolve on a planet immersed in a wholly hostile environment.

The *first* is the obvious drop in air pressure as one rises higher and higher above the surface of the Earth, and with it, a rapidly dwindling supply of oxygen to breathe. The *second* is more direct exposure to the full intensity of solar electromagnetic radiation, from the shortest to the longest wavelengths but particularly the highly energetic and penetrating extreme-ultraviolet and x-ray radiation from the Sun, which we never feel on the surface of the Earth. The *third*—most dangerous to life and health and hardest to avoid—is a direct exposure to streams of highly energetic atomic particles that arrive from solar disturbances and cosmic sources.

The Ocean of Air

Like fish in water, we can survive without artificial aid only at the bottom of an oxygen-rich ocean of air. Our ability to ascend safely above this abyssal depth—to the top of the atmosphere or even part way there—will always require that we bring with us our own pressurized atmosphere.

In the range of altitudes at which jet aircraft fly—seven or eight miles above the surface—air pressure and available oxygen are depleted by factors of four to six compared to conditions at the level of the seas. There—as in spacecraft operating at heights where ambient pressure and available oxygen have fallen by factors of more than a million—conditions essential for life are met by pressurizing and continually replenishing the air within the aircraft cabin.

On those occasions when astronauts venture *outside* their hermetically-sealed spacecraft—as during EVAs or on the surface of the Moon or Mars—the necessary environment is reproduced by pressurizing their air-tight space suits, helmets and gloves to atmospheric levels, and by carrying with them a portable oxygen supply.

Enhanced Ultraviolet and X-Ray Radiation

The threat from solar ultraviolet and x-ray radiation—which increase in intensity the higher we go—is more easily kept at bay.

Near-, intermediate- and extreme-ultraviolet radiation is able to penetrate the skin, damaging both the epidermis and the living tissue beneath it. Ultraviolet radiation is also particularly damaging to the eyes, and to the immune system. The aluminum body of an aircraft or spacecraft is more than adequate to block the full spectral range of the ultraviolet, as is the material of which windows in either of these are made. When outside the spacecraft, an astronaut's space suit, helmet and highly-reflective visor provide an equivalent level of protection against direct solar ultraviolet radiation.

Solar X-Rays

Solar x-rays, which are considerably more energetic than those employed in medicine, dentistry and airport security systems, are more hazardous than ultraviolet radiation and not as easily blocked.

As has been known and put to use for more than 100 years, clothing and flesh are transparent to x-rays. This is also true of much denser materials when directly exposed to the more energetic x-rays that come from the Sun. It is also known that a sufficiently long or cumulative exposure to x-rays—and particularly the more energetic—can damage and alter the further division of human cells, in the same way that energetic particles do: by ionizing some of the atoms of which they are made.

When within the metal shell of a spacecraft, astronauts are adequately shielded from both far-ultraviolet and x-ray radiation. The risk comes during extra-vehicular activity or when an astronaut in a space suit permeable to x-rays finds herself or himself on the sunlit surface of the Moon or Mars at a time when a major solar flare erupts. Unless cover is immediately found, the time of direct exposure to x-rays could last up to an hour or more: as compared to the fraction of a second employed in medical and dental x-rays. Moreover, the area potentially exposed is more extensive: not one's jaw or lungs or forearm, but everything there is from head to toe.

A Sun Intensely Bright

One should never look directly at the face of the Sun, unless it is severely dimmed in some way or seen through a very dark and dense optical filter.

When seen from above most or all of the ocean of air, the potential risks are even greater.

In space, where there is no atmospheric absorption or scattering of sunlight, the disk of the Sun is definitely brighter, and much more so at the violet end of the visible spectrum. It was with this in mind that the helmets worn during EVAs—and by *Apollo* astronauts on the Moon—are equipped with highly-reflective, gold-mirrored visors.

There is as well the potential of a kind of snow-blindness, if anyone *within* a spacecraft should catch an inadvertent glimpse at the first rays of the Sun at sunrise: an event that is repeated about every ninety minutes, or sixteen times each day for those in lower orbits about the Earth. And each time—as on the road to Mandalay—the dawn comes up like thunder.

In truth—as with many other familiar things—there is no dawn in space. Instead, the first sliver of the rising Sun appears abruptly and without warning, for in the absence of an atmosphere there is no gradual bluing of the coal-black sky or any colored glow on the horizon to herald its coming, and to condition the dark-adapted eye. Moreover, when it appears, the Sun also climbs above the horizon far faster—as in a speeded-up movie—since the apparent motion of the Sun through the sky is, when in orbit about the Earth, driven not by the 24-hour rotation of the planet but the 1½ hour orbit of the spacecraft around it.

For an astronaut who—while looking out, perhaps, at the black starlit sky—chances to see the first bright edge of the fast-rising Sun it would seem like the shock of a flash-bulb in a totally darkened room, though in this case as bright as a welder's torch.

Solar Energetic Particles and Cosmic Rays

Because of their mass and extremely high speeds, atomic particles carry very high energies and are therefore far more lethal than the damaging ultraviolet or x-ray electromagnetic radiation that comes from the Sun. They are also much harder to protect against, for not only space suits and helmets but the materials from which spacecraft are made present few obstacles to these speeding bullets.

Astronauts have even "seen" some of these invasive high energy particles while in space. Beginning with the first *Project Mercury* flights, astronauts in orbit about the Earth have reported seeing flashes of bright light, which are visible whether their eyes were open or closed. These were soon identified as the tracks of cosmic rays or solar energetic protons that produced what is known as Cêrenkov light on their way through the astronaut's head: more specifically, through the viscous medium that fills the eyeball. Cêrenkov radiation, which has been known since 1939, is a form of light created by charged atomic particles when they pass through a transparent substance at a speed greater than the velocity of light in that medium.

Particles from solar flares or cosmic sources with energies from ten to several hundred million electron volts (MeV) can easily penetrate the thin metal skin and frame of a spacecraft: which, as in airplanes, is designed with minimum weight in mind. A 100 MeV proton traveling at velocities close to that of light itself can pass through a good 1½ inches of aluminum—about three times the thickness of the steel hull of the Titanic—before it is stopped. Lighter and hence weaker 3 MeV electrons—of the sort that swarm in great numbers within the Earth's radiation belts—can make it through a sheet of aluminum a quarter-inch thick. And the most energetic particles—galactic cosmic rays (GCRs) with energies as great as a billion or more electron volts (GeV), are for most situations in space essentially unstoppable.

Thus the health hazard of greatest concern in manned spaceflight, as in high-altitude jet aircraft travel, is exposure to high-energy solar atomic particles and galactic cosmic rays.

The expected dosage in a high altitude aircraft—even over the poles or during major solar flares—is much reduced, since by the time these speeding particles (or more correctly, their descendants) reach the altitudes at which conventional aircraft fly, they have lost much of their original prodigious energy through encounters with atoms and molecules of air on their way down.

This is not the case for spacecraft. What is more, the expected dosage increases considerably with travel in higher orbits that take them into or near the Earth's radiation belts: and more so beyond the protective arms of the magnetosphere.

Another hazardous zone for spacecraft that travel in Earth orbits within or below the magnetosphere is the South Atlantic Anomaly, or SAA: a portion of the Earth's magnetic field located over southern South America and the South Atlantic Ocean where the field strength is considerably reduced. In this danger zone—which might seem reminiscent of the Bermuda Triangle were it not in this case real—energetic particles in the Earth's inner radiation belt are able to penetrate more deeply into the thermosphere, to altitudes where so many spacecraft in near-Earth orbits fly.

FLIGHT PATHS OF AIRCRAFT AND SPACECRAFT IN ORDER OF INCREASING EXPOSURE TO HIGH-ENERGY PARTICLES AND RADIATION

FLIGHT PATH AND ALTITUDE	EXAMPLE FLIGHT VEHICLES	HAZARD	EXACERBATING FACTORS Defined in the Legend Below
PRESENT DAY JET AIRCRAFT [6 - 10 miles altitude]	Commercial Airlines, Corporate and Military Aircraft	Cosmic Rays, Solar Protons	Polar Routes, CMEs
NEXT GENERATION JET AIRCRAFT [10 - 12 miles]	Commercial airlines, military aircraft	Cosmic rays, solar protons	Polar Routes, CMEs
EARTH ORBITS BELOW THE MAGNETOSPHERE: Low Earth Orbits (LEO) [150 - 500 miles] at low latitudes	Space Shuttle, Hubble Space Telescope (375 mis)	Cosmic rays, solar x-rays and protons, trapped particles	SAA, EVAs, CMEs, MAG STORMS
LOW EARTH ORBITS (LEO) at higher latitudes [150 - 500 miles]	International Space Station	Cosmic rays, solar x-rays and protons, trapped particles	SAA, EVA, CMEs, MAG Storms
EARTH ORBITS IN CLOSE PROXIMITY TO THE LOWER RADIATION BELT: Medium Earth Orbits (MEO) [16,000 miles]	Fleet of GPS Spacecraft	Trapped particles, solar x-rays and protons, cosmic rays	MAG Storms, CMEs
SPACE TRAVEL WITHIN THE RADIATION BELTS: GeoSynchronous Orbits (GEO) [22,200 miles]	Telecommunications Satellites; Manned Spacecraft en route to the Moon or Mars	Cosmic rays, solar protons, trapped particles	MAG Storms, CMEs
SPACE TRAVEL BEYOND THE MAGNETOSPHERE: Enroute to, or on the surface of the Moon	Apollo Spacecraft and Lunar Modules; Planned Missions to the Moon	Cosmic rays, solar x-rays and protons	Duration, EVAs, CMEs
ENROUTE TO, OR ON THE SURFACE OF MARS	Projected Human Exploration of Mars	Cosmic rays, solar x-rays and protons	Duration, EVAs, CMEs

LEGEND
CMEs   During or following a major solar flare or CME
EVAs   During extra-vehicular activity of any kind
MAG   Occurrence of a major geomagnetic storm
Polar Routes Aircraft flights that cross polar or high latitude regions
SAA   Passage through the South Atlantic (magnetic) Anomaly
Duration   Manned space missions of long duration

Among the affected are the *Space Shuttle*, the *International Space Station*, and a large number of other unmanned spacecraft. As an example, one of the instruments on the *Hubble Space Telescope* was found to be so disrupted in passage through the SAA that it was routinely switched off during each of the Hubble's brief passages through this zone.

The Physiological Effects of Ionizing Radiation

The risks of exposure to very energetic atomic particles lie in their ability to tear electrons from (or ionize) atoms in cells that make up human tissue. Because of this, cosmic rays, solar energetic particles and even solar x-rays are often described as *ionizing radiation* (to distinguish their effects from those of solar *electromagnetic radiation* in the infrared, visible, ultraviolet portions of the spectrum.) They can do this whether or not the impacting particles carry an electric charge.

The ability of speeding atomic particles to strip electrons from atoms in living tissue depends chiefly on their energy. In terms of their effect on life, the chargeless neutrons that are released as secondary cosmic rays pose as great a hazard as particles of equal energy that carry a positive or negative charge.

When neutral atoms in living cells are ionized, the chromosomes and DNA within them are altered in ways that can lead to cellular mutations and the risk of cancer.

Exposure to extreme doses of ionizing radiation can produce other, more immediate effects: skin burns which are slow to heal, cataracts, nausea, vomiting, damage to the nervous system, and depletion of the immune system. In short: the symptoms of radiation sickness experienced by cancer patients receiving radiation therapy, or more severely by the many victims of atomic bomb explosions over Hiroshima and Nagasaki some sixty years ago.

The Importance of Dosage

The extent to which those in exposed aircraft flights or in space are ultimately affected is determined by both the *amount* of radiation received at any time (as during a medical x-ray, or a half hour's exposure to solar energetic particles) and the *cumulative total* of ionizing radiation encountered in the course of one's life.

⊙   ⊙   ⊙

We all receive some ionizing radiation every year, no matter where we live or what we do. Should we undergo medical or dental x-rays, we are exposed to a brief and localized dose. More significant in most cases is an ever-present dose of ionizing radiation from some of the rocks and minerals and soils that are beneath and all around us, from which in the course of a year it is possible to accumulate a dosage of up to ten medical x-rays.

Aircraft crews and passengers receive a small amount of potentially-damaging radiation in every flight that crosses polar and sub-polar regions, which include most non-stop flights to and from Asia from northern American airports and many to destinations in northern Europe. Astronauts are exposed to greater levels from cosmic rays and other energetic particles each time they venture into space.

The recommended *maximum annual dosage* for anyone is equal to about ten medical x-rays. A round-trip flight over a polar route between New York to Hong Kong will on average accrue an exposure equivalent to about two of them. In space the dosage increases dramatically: from cosmic rays, solar energetic particles and from energetic particles trapped in the Earth's radiation belts. The total dosage is a function of the flight path, the time spent in EVAs, the state of solar activity, and the overall duration of the space flight.

⊙ ⊙ ⊙

Almost all manned space flights, including the relatively short flights of the *Space Shuttle* and the far longer stays on the *Space Station*, have remained at altitudes of no more than about 300 miles above the surface: well below the Earth's radiation belts and also, for most solar particles, within the protection of the Earth's magnetosphere.

Under these circumstances the cumulative exposure to ionizing radiation of those on board—though much greater than what we receive on the ground or the air—has been shown to be of little consequence, and as yet, no cause for alarm. For example, exposure to cosmic rays during a record one-year stay in low Earth orbit aboard the *Mir* spacecraft was estimated to increase the risk of future cancer for cosmonauts on board by about one percent. Experience has shown that under ordinary solar conditions, orbital flight below the level of the radiation belts and magnetosphere are not a major health concern.

However, for a manned spacecraft traveling *beyond* the magnetosphere during a particularly intense solar flare, or worse, during a long-lasting CME particle event, the ionizing radiation received each hour within the spacecraft can equal that of six or seven *thousand* chest x-rays, with even greater consequences if caught outside, on EVA.

This has yet to happen, but it could have and almost did during the five-year period of NASA's manned *Apollo* missions to the Moon: eleven flights, each lasting from ten to fourteen days, which for the first time carried human cargo beyond the protection of the magnetosphere and into the full force of solar and cosmic particles.

The Disaster That Almost Happened

The first, last and only space ventures of this kind were the nine *Apollo* missions conducted in late 1968 through early 1972, each of which completed the 480,000 mile voyage to the Moon and back.

Off and on in this 51-month period—which followed the peak of a moderately strong sunspot cycle—25 astronauts, on flights spaced on average about six months apart, lived and worked in deep space, shielded only by the *Apollo* command module spacecraft and the smaller and the less shielded lunar lander.

MANNED SPACE FLIGHTS THAT HAVE GONE BEYOND THE MAGNETOSPHERE

MISSION	CREW	LAUNCH DATE	TIME SPENT IN DEEP SPACE	TIME SPENT ON THE MOON
Apollo 8	F. Borman J. Lovell W. Anders	1968 Oct 11	6 days	None
Apollo 10	E. Cernan J. Young T. Stafford	1969 May 18	8 days	None
Apollo 11	N. Armstrong M. Collins E. Aldrin	1969 July 16	7 days	21 hrs, 36 mins
Apollo 12	C. Conrad R. Gordon A. Bean	1969 Nov 14	10 days	31 hrs, 31 mins
Apollo 13	J. Lovell F. Haise J. Swigert	1970 Apr 11	6 days	None
Apollo 14	A. Shephard S. Roosa J. Mitchell	1971 Jan 31	9 days	33 hrs, 31 mins
Apollo 15	D. Scott A. Worden J. Irwin	1971 July 26	12 days	66 hrs, 55 mins
Apollo 16	J. Young* T. Mattingly C. Duke	1972 Apr 16	12 days	71 hrs, 2 mins
Apollo 17	E. Cernan* R. Evans H. Schmitt	1972 Dec 7	13 days	75 hrs

*Crewed Apollo 10 as well

In all, these nine bold missions kept astronauts immersed in this unknown and potentially hazardous environment for a total time of almost three months. In terms of possible major impacts from the Sun, the one to two-week missions placed these deep-space travelers in harm's way for about 6% of the time during a 51-month period of moderate activity in the life of the Sun: the intervals depicted with blue diamonds in the following chart.

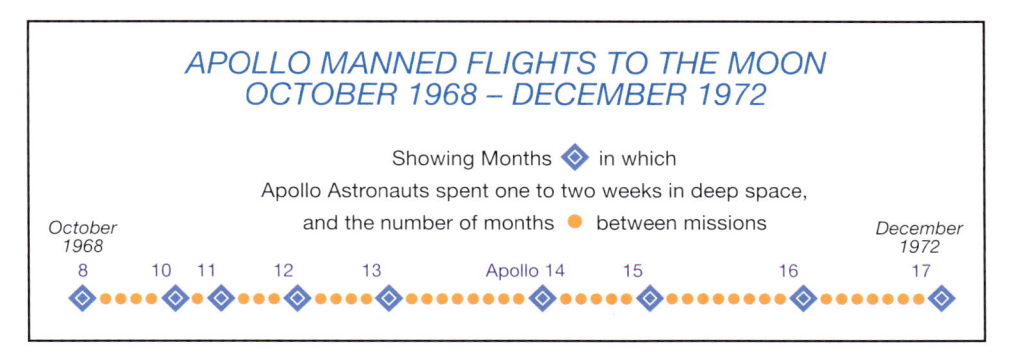

APOLLO MANNED FLIGHTS TO THE MOON
OCTOBER 1968 – DECEMBER 1972

Showing Months ◆ in which
Apollo Astronauts spent one to two weeks in deep space,
and the number of months ● between missions

October 1968 — 8 10 11 12 13 Apollo 14 15 16 17 — December 1972

In addition, for a total time of 12½ days men lived and worked on the wholly unprotected surface of the Moon. And although they never wandered far from the lunar lander, it would have offered no protection at all against extremely energetic solar particles that rained down on the surface of the Moon. The only safe place would have been against a shaded, sloping wall of a lunar crater, were such a refuge close at hand. Even then, a place in the shade would not offer full protection against incoming charged particles from the Sun, since the earliest particles arrive along curved paths and the others from all directions (other than up from the ground), as a result of scattering by magnetic fluctuations encountered en route from the Sun.

While many smaller flares and unseen CMEs occurred during the period of the *Apollo* flights, there was but one anomalously-large solar proton event.

⊙ ⊙ ⊙

In early August of 1972, well into the declining phase of solar cycle 20 and about midway in the four-month down time between the *Apollo 16* and *17* missions, the Sun displayed an abrupt and unexpected increase in solar magnetic activity. The result was ten days that shook the world: intense solar flares and CMEs, one after another. From this prolonged barrage came a steady stream of solar energetic protons (SEPs) that provoked magnetic storms, auroral displays, and severe disruptions of radio communications and electric power transmission.

Solar flares most often last for a matter of minutes followed by more extended particle showers that arrive at Earth several days later. But the super flare that was seen on the Sun in the early morning of Monday, August 7, 1972 lasted more than *four hours*, making it one of the largest and most intense of any that have ever been observed. In the same 10-day period CMEs showered the Earth for extended periods of time. This combination of intense flares and CMEs exposed our planet in those few days in early August to record-breaking levels of high-energy particles.

Solely by chance, the *Apollo 16* mission had returned safely from the Moon about four months *before*, on April 27; and the next and last of the series, *Apollo 17*, left for the Moon about four months *later*, on December 7.

Had there been a significant delay in *Apollo 16*, a moved-up *Apollo 17*—or had the Sun erupted as severely but a few months earlier or later—the three astronauts on board would have been exposed for hours or even days to potentially lethal ionizing radiation, whether they were at the time within the command module, the less shielded lunar lander, or on the surface of the Moon.

☉ ☉ ☉

The heaviest dose during this period would have come principally from long-lasting streams of energetic protons from CMEs, combined with the shorter exposure to similar particles from the August 7 flare. During the first half-day of their exposure to atomic particles from that one event, the *Apollo* astronauts would have exceeded the recommended maximum *yearly* dosage of radiation to the eyes and internal organs and the maximum *lifetime* dosage to the skin. As a result of this exposure they would almost certainly have experienced severe symptoms of radiation sickness, which for some or all of the three adventurers could have proven fatal.

The initial blast of high-energy protons from so intense a flare would have reached the spacecraft and the surface of the Moon within twenty minutes after the visible explosion was seen at solar observatories around the world. But the actual warning time for the *Apollo* astronauts would have been reduced to less than twelve, because of the time it takes for the light that announced the flare to reach the Earth. Moreover much or all of this would have been spent in human reaction and response times, and in communications among solar observers, flight controllers, and the spacecraft crew.

In these few minutes what must be evaluated is first the relative intensity of the flare; second, whether particles expected from it are likely to strike the Earth-Moon vicinity; and third, the nature of the risk to *Apollo* astronauts and the mission itself.

At the very most the astronauts might have had a few short minutes to find whatever cover they could, before the first wave of particles came upon them. In the command module—or worse, the lunar lander—there would have been no place to hide from the most energetic solar particles. Were they working on the surface of the Moon when the urgent warning was received, they would have found only limited protection from the first-arriving particles were they to hide behind a shaded wall of a lunar crater, for reasons cited earlier. But this is not the case for the many others that follow the initial blast. These particles, which can arrive over many hours or days in the case of CMEs, are deflected or scattered by fluctuations in the heliosphere and hence arrive from all directions in the sky.

☉   ☉   ☉

Presently-planned round-trip flights to the Moon, probably lasting about two weeks, and the far longer stays envisioned in the recent decision to establish a permanent base there will again carry the potential of severe hazards from ionizing radiation: because of the period of time spent outside the Earth's atmosphere and magnetosphere; the finite probability in most seasons of the solar cycle of intense flares or CMEs; and the engineering challenges involved in providing adequate shielding against GCRs and the most energetic solar protons while in deep space.

Envisioned trips to Mars, lasting from two to three years and with stays of up to eighteen months on the exposed Martian surface, present problems far more severe. This is in part because astronauts would be exposed to the full fury of the Sun not 6% of the time, as was the case in the four years of *Apollo* flights, but *all* of the time through a continuous period of three to four years, in addition to the continuous exposure to very high energy galactic cosmic rays.

The greater risk to life and health on any *extended* mission beyond the Earth's atmosphere will be the accumulated dosage of galactic cosmic rays that are more energetic by far than solar particles and which arrive from everywhere, every day and around the clock. In addition, on excursions as long as these—50 to 75 times longer than those to the Moon—the voyagers are almost certain to experience at least one and more likely several major solar flares, and countless CMEs that accelerate high-speed solar wind particles to very high energies.

☉   ☉   ☉

To avoid the certain hazard of prolonged cosmic ray exposure—and to reduce the odds of a solar catastrophe—long missions, particularly, will require a practical solution to the yet-unsolved problem of providing adequate shielding. The first that comes to mind is probably that of sheathing the spacecraft living quarters in heavy metal many inches thick, or a blanket of water at least 15 feet deep with a mass of more than 500 tons, which is nearly 20 times what the present space shuttle can carry. Technical and weight considerations also pose severe engineering challenges for the notion of containing the spacecraft in its own protective magnetic field, which would need to be more than half a million times stronger than the Earth's field for shielding against GCRs.

Other options, now under study or consideration include the development of much larger rockets that would reduce exposure by speeding the journey to and from the Moon or Mars, and a proposal to enclose the inhabited portion of the manned spacecraft within a protective shell formed by a wrap-around cluster of long cylindrical tanks filled with liquid hydrogen. The nucleus of each hydrogen atom is a single proton, which is a relatively efficient absorber of some of the kinetic energy of particles—like cosmic rays—that happen to collide with it. By this scheme, the heavy cargo of liquid hydrogen fuel that must be taken along to lift the returning spacecraft from the surface of Mars would serve an additional, second purpose: as a partial buffer to protect against high energy cosmic rays during both the long outward voyage and the even longer stay on Mars. But all of these are as yet untried.

The serious threat of lengthy exposures to high energy cosmic rays—and the difficulty of shielding against them—may indeed make it more prudent to schedule extended manned missions like those to Mars during years of *high* rather than low solar activity. The bases for this are that when the Sun is most active the flux of galactic cosmic rays drops by 20% or so; and the fact that particles accelerated by CMEs and from flares—while considerably more prevalent then—are less energetic and therefore relatively easier to shield against.

The hazards of lethal cosmic rays may prove to be the principal obstacle to extended manned missions of any kind, including long-envisioned human exploration and colonization of the cosmos.

Impacts on Spacecraft, Space Equipment and on Observations of the Earth From Space

All things launched into space enter a hostile world. When a spacecraft reaches the exosphere—at an altitude of about 300 miles—it is directly exposed to EUV and x-ray radiation. As it continues its upward path—first within and then beyond the magnetosphere—it is more fully exposed to high energy solar particles; to the piercing rain of even more energetic atomic particles from distant cosmic eruptions; to violent gusts in the solar wind; and to sudden and potentially catastrophic bursts of highly energized atomic particles from solar flares and CMEs: the hurricanes and tsunamis of the heliosphere.

There are today some 850 spacecraft operating within this exposed environment, put there at a total cost of more than 100 billion dollars to serve military, civic, scientific, and commercial needs. The U.S. now owns or operates slightly more than half of these.

About 100 are there to support national security, surveillance, and military operations. A similar number gather unclassified data on all aspects of the Earth and space for the various federal agencies, including NOAA's weather and climate services, NASA, and the Departments of Agriculture and the Interior. But by far the largest fraction fall in neither of these two categories, but within the commercial sector, serving the burgeoning satellite communications industry.

Interruptions, or significant damage of any kind to this combined fleet of spacecraft can directly affect many aspects of modern life, and they happen all the time. Among the impacted areas are meteorological, oceanographic and geophysical observations taken from instrumented spacecraft, including the pictures of clouds and weather systems seen on cable or the nightly news; satellite telephones and most telephone land lines; geographic positioning systems; the relayed signals that allow broadcast television and satellite radio, police and emergency communications and the transfer and rapid exchange of commercial information involving transactions such as credit card purchases, stock exchanges, and automated teller machines; marine and aircraft navigation and aircraft traffic control; police and emergency communications; and a host of national security functions, including the minute-by-minute tracking of satellites and space debris.

⊙ ⊙ ⊙

Bursts of energized electrons, protons and heavier ions from solar flares and CMEs are fully capable of damaging or disrupting the operation of many common components of space equipment. Among them are circuit boards, computers, computer software and storage devices, electrical and electronic cabling, solar cells, and photo-sensors. The same can happen when streams of high-speed solar wind provoke geomagnetic storms; when the spacecraft finds itself near or within the Earth's radiation belts; or from the continuous barrage of high-energy cosmic rays.

As noted earlier, the cost-driven shift from custom-produced "hardened" components to those bought off the shelf has greatly increased the vulnerability of spacecraft and space equipment to solar events. As has the trend toward miniaturization of computers and other electronic instruments.

Individual components of computer and other micro-electronic equipment are—unlike our own bodies—vulnerable to single particle events, with immediate impacts on the system in which that element is a part. In contrast, the primary concern in manned space flight is not so much individual events but the *accumulated dosage* of ionizing radiation.

Times of Particular Hazard

Because flares and the fastest CMEs occur more often when the Sun is more active, the most vulnerable times for spacecraft and space equipment—whether in orbit around the Earth or far beyond it—are the four or five years during and following the peak of the 11-year cycle, when solar eruptions sufficiently large to disrupt satellite operations will occur as much as 15% of the time. The effects of satellite drag are also greatest in years of high solar activity.

At the same time, while far more frequent when the Sun is more active, there is no respite from CMEs, which occur throughout the solar cycle. What is more, high-speed streams of plasma in the solar wind, which can accelerate charged particles in the Earth's radiation belts, are more prevalent when the Sun is *less* active. Because of this, potential damage to spacecraft and space equipment, though greater in years of maximum activity, is possible in all phases of the solar cycle.

Spacecraft that travel beyond the protection of the upper atmosphere and magnetosphere face an abrupt jump in the number of potentially-damaging solar particles when CMEs and flares occur. For those within the protection of the magnetosphere—where most spacecraft operate—these impulsive solar events can exert a similar effect, either directly or by increasing the numbers of energetic particles held within Earth's radiation belts.

CMEs also reshape the form of the magnetospheric shield, pushing it closer to the Earth on the Sun-facing side and blowing the cover, in a sense, for spacecraft in higher, geosynchronous orbits at altitudes of about 22,200 miles. These and high-speed solar wind streams can also provoke magnetic storms by accelerating particles in the Earth's inner radiation belt.

Flight Paths of Greatest Risk

The level of risk to spacecraft and space equipment is very much dependent where they operate. The most hazardous are trajectories that take it (1) into or near the radiation belts; (2) beyond the Earth's magnetic field; (3) in polar orbits; or (4) in any inclined, equatorial orbit that passes through either sub-polar regions or the South Atlantic Anomaly, where the Earth's field is notably weaker.

By far the most hazardous are those that spend time in the heart of the Earth's radiation belts or leave the protections of the Earth altogether. In addition to past and planned manned missions to the Moon and possibly to Mars, these include the many unmanned spacecraft that have been sent out to explore and monitor conditions in near-Earth space, and those commissioned as remote explorers of the Moon and planets and other objects within and beyond the solar system.

Spacecraft at the Lagrangian Points of the Sun-Earth System

Other instrumented spacecraft operate in high-Earth orbits that take them far beyond the outer periphery of the magnetosphere. The *Solar and Heliospheric Observatory*, or *SOHO*, has for more than ten years circled the Sun at a fixed distance from the Earth of about 940,000 miles—about 1% of the distance to the Sun—while always remaining on the Sun-Earth line.

The *ACE (Advanced Composition Explorer)* satellite, which samples conditions in the solar wind and gives advance early-warnings of conditions within approaching streams of solar particles, is another spacecraft that operates at this unique and faraway location: a place of neutral gravity between the Sun and the Earth-Moon system known as the L1 Lagrangian point, in honor of its predictor, the French mathematician Joseph-Luis Lagrange (1736-1813).

It was Lagrange's solution to a theoretical problem in mathematics that identified five unique "neutral points" in the combined gravitational fields of the Sun and the Earth-Moon system where a body should experience no net gravitational force. The most accessible of these, today known as L1, was empirically tested and put to practical use for the first time 184 years after his death.

At this place and distance the pulls of gravity from the closer and lighter Earth and Moon are neutralized by the gravitational tug in the opposite direction from the massive but more distant Sun. Here a spacecraft is in essence weightless, neither drawn back to the Earth nor from it toward the Sun. And with no place to fall, if brought to a stop it will remain in that vicinity: as though fixed to a rigid spoke that connects the center of the Earth to the Sun.

The pulls of gravity at the Lagrangian points are not perfectly stable, however, due to the non-circularity of both the orbit of the Earth and that of the Moon. As a result, spacecraft such as *SOHO* and *ACE* trace out small orbits of their own, with periods of about six months, that are centered on the L1 point. The cost of these orbital imperfections on the operation of spacecraft residing there is a need to spend on-board fuel to make minor corrections in their position, in order to keep them on station, lest they drift beyond the L1 zone into their own independent orbit around the Sun.

A distinctive set of more than 200 minor planets called the Trojan asteroids are among other astronomical bodies that have long been affected by the mathematical singularities that Lagrange discovered. In the asteroid case the two objects whose gravitational attractions balance each other at five Lagrangian points are the Sun and Jupiter. The Trojan asteroids circle the Sun like Jupiter, in roughly 12-year orbits, held in the vicinity of the L1 Lagrangian point in this *other* three-body system. They too are made to oscillate in small circles about that mathematical point, due in part to the gravitational disturbances of the many moons of Jupiter.

Polar Orbits and the South Atlantic Anomaly

When spacecraft in polar orbits pass through regions where open field lines extend outward from the magnetic poles of the Earth they cross a region where potentially-damaging atomic particles make their way down into the upper atmosphere. Also in harm's way are those in inclined equatorial orbits that carry them through the South Atlantic Anomaly, where an enfeebled magnetic field allows the Earth's inner radiation belt to expand downward into their range of altitude.

Geosynchronous and Geostationary Orbits

The most vulnerable of flight trajectories that are kept *within* the protective arms of the magnetosphere are the geosynchronous and geostationary orbits at an altitude of 22,200 miles that keep spacecraft above a specific region on the Earth's surface. Included in these is the ever-growing number of commercial telecommunications satellites, as well as many scientific and other payloads. Under most conditions, spacecraft at this altitude operate at the outer edge of the outer radiation belt.

In contrast, the *safest* of all flights in space—in terms of potential impacts of high energy particles on spacecraft, equipment and spacecraft operation—are shorter missions that operate in low-Earth orbits (at altitudes of 120 to 300 miles) during years of minima in the 11-year solar cycle. An example was the historic, first flight of John Glenn, who circled the Earth, alone, in a *Project Mercury* capsule at low latitudes for six hours on the 20th day of February in 1962, well into the declining phase of solar cycle 19.

ORBITAL ALTITUDES OF SPACECRAFT IN RELATION TO PROXIMATE ATMOSPHERIC FEATURES

FEATURE / SPACECRAFT ORBIT	DISTANCE ABOVE THE EARTH'S SURFACE IN MILES	DISTANCE FROM CENTER OF THE EARTH IN EARTH RADII
Bottom of F Region in Ionosphere	90	1.02
Height of Auroral Displays	170 – 360	1.04 – 1.09
Low-Earth Orbits (LEO)	≤ 1200	≤ 1.3
Space Shuttle / International Space Station / Hubble Space Telescope	250	1.06
Top of F Region	650	1.2
Medium Earth Orbits (MEO): GPS Transmitters	6000 – 16,000	2.5 - 5
Inner Radiation Belt	650 – 12,000	1.2 – 3.0
Gap between Inner and Outer Belts	12,000 – 16,000	3 - 4
Outer Radiation Belt	16,000 – 24,000	4 – 6 or more
Geosynchronous Earth Orbits (GEO): Weather Satellites / Telecommunications Satellites	22,200	6.6
Spacecraft Operating at L1 Lagrangian Point: SOHO (Sun Imaging Spacecraft) / ACE (First Sentinel Spacecraft)	930,000	234

Destructive Particles From the Sun and the Earth's Radiation Belts

As with projectiles fired from a gun, the most destructive of atomic particles are the largest, heaviest and fastest moving.

Protons, which are about 2000 times heavier than electrons, are the most common ions released from solar flares and CMEs. When accelerated to speeds that approach half the velocity of light—which happens in major flares and CME-driven shocks—they are hard to stop and extremely destructive.

When an atom of hydrogen (consisting of one proton, one electron and nothing else) is ionized, its single electron is torn away from the positively-charged proton to which it was bound. Since hydrogen is far and away the most abundant element in the Sun—90% by number, 70% by weight—it is no surprise that protons comprise so large a fraction of the energetic particles thrown off from the Sun, or that they are the leading agents of disruption and damage to spacecraft and space equipment.

Heavier solar particles, including ionized helium (alpha particles) or the electron-stripped atoms of other elements, are also ejected. The *heaviest* of charge-accelerated solar particles—the cannonballs of solar projectiles—are ionized atoms of iron, which, though their speeds are not as great, can be more than fifty times heavier than protons.

Other potentially-damaging particles come from the hordes that are trapped and held in the Earth's radiation belts. Some of these came from the Sun, in the solar wind or from violent solar eruptions.

But mixed among them are an equal or greater number that entered the radiation belts not from other worlds but from our own: migrant electrons, protons and heavier ions from the ionosphere and thermosphere that diffused upward, and because of their electrical charge, were captured and held in the magnetosphere. The disturbances that set any of these charged particles free are geomagnetic storms triggered by solar eruptions or streams of high-velocity solar wind that for a brief time disrupt the bonds that held them there.

Cosmic Rays

Galactic cosmic rays are the most energetic of particles found in space. The most prevalent are high-energy protons, but as with particles from the Sun, there are also much heavier ions, ranging from the nuclei of helium atoms (atomic weight four, or four times the weight of a proton) to the relatively abundant ions of iron (weight 56). And they travel outward at speeds that can considerably exceed most particles of solar origin.

The reason for the great energies and high speeds of cosmic rays, and why they include heavy ions such as iron, nickel and zinc, is that they are propelled outward from cataclysms far more violent than any large solar flare—or a thousand of them set off all at once. The source of most cosmic rays is thought to be the explosion—long ago and far away—of an entire star, called a supernova: as though the star we live with were one day suddenly and unexpectedly blown to bits.

A cosmic catastrophe of this scale releases an almost unimaginable amount of energy all at once, propelling extremely energized electrons, protons and heavier ions at phenomenal speeds in all directions through all of space. Because of their high energies, it is nearly impossible to shield a spacecraft, a computer, other space equipment, and any human passengers against them, since those of highest energies can pass through many inches of heavy metal. They are also nearly impossible to avoid, for cosmic rays are everywhere throughout the universe, and arrive in the solar system—but little slowed after tens of thousands of years of travel—in a cross fire of killer rays, coming from all directions.

Atmospheric Drag

Spacecraft operations are also directly affected by the two- to ten-fold increase in solar EUV and x-ray radiation in years when the Sun is more active.

When radiation in this short-wave region of the solar spectrum is absorbed by atoms of air in the thermosphere it raises the temperature there. In response, the thermosphere swells, expanding upward into regions of the more diffuse exosphere where spacecraft in low-Earth orbits fly.

This intrusion of denser air from the thermosphere increases the friction—or drag—on spacecraft that travel through it. The added drag slightly slows the space vehicle, causing it to drop to a lower altitude. There the air is even more dense, leading to a compounding effect on spacecraft altitude.

Unless the spacecraft is boosted to a higher operating altitude—at the cost of some of its limited supply of orbital control fuel—this chain of events will inexorably pull the spacecraft down into the far denser middle and lower atmosphere of the Earth, where increased friction will ultimately tear it apart.

As the thermosphere expands, the rising tide affects all boats that orbit the Earth below an altitude of about 1000 miles. The *Hubble Space Telescope* and *Space Station* are among many spacecraft that have made orbital corrections using control jets to compensate for the changing moods of the Sun: as they soon will again, as solar activity begins its climb toward a maximum of solar cycle #24 in about 2013.

These thermospheric expansions occur in response to both the year-to-year cyclic rise in solar activity and—on scales of hours to days—the intense bursts of short-wave radiation from major solar flares.

The increased atmospheric drag that accompanied the Bastille Day flare of July 14, 2000 led to the immediate demise of Japan's *Advanced Satellite for Cosmology and Astrophysics (ASCA)* by disabling its ability to maintain its orientation in space. This sent it into a catastrophic spin that led to a total loss of power, since the spacecraft could no longer keep its solar power panels directed toward the Sun.

☉   ☉   ☉

Skylab, the first manned space station, fell victim to the very object—the Sun—which it had been built and launched to observe, as a result of a year-by-year increase in thermospheric drag.

Carrying a battery of six of the most advanced telescopes ever pointed at the Sun, *Skylab* was lofted into low-Earth orbit in 1973, during the declining phase of solar cycle #20. Through the efforts of ground controllers and three successive teams of three astronauts—who relieved each other in space at the end of prolonged stays—the spacecraft and its crews completed its ambitious mission within eight months.

On the 8th of February 1974 the third, last and longest-staying crew turned out the lights, closed the door and returned to Earth in the space capsule that had brought the first team there. The abandoned spacecraft—as big as a boxcar and weighing about as much—was left behind to circle the Earth: a ghost ship like the *Mary Celeste*, which was found adrift in mid-ocean in 1872, fully laden and with no one on board.

In 1975, magnetic activity on the surface of the Sun began an extremely steep rise, climbing to the unusually high maximum of solar cycle #21 that was reached three years later, in 1978. The steep increase in solar activity brought an ever-increasing dose of short-wave radiation from the Sun, swelling the thermosphere and dragging the abandoned *Skylab* farther and farther down into the atmosphere: much faster than what had been anticipated.

In 1979, two years before the completion of the first *Space Shuttle*—which, had it been ready, could have been sent to boost the sinking spacecraft into a higher and safer orbit—the abandoned *Skylab* spiraled downward, out of control into the dense air of the lower atmosphere. There the most sophisticated and expensive solar observatory in the world broke apart into chunks, large and small, which fell to the ground: by good fortune over an uninhabited stretch of the Australian desert.

Impacts on Micro-Circuits and Computer Systems

High levels of electrostatic charge—created through contact with energized electrons—accumulate on the outer surface of a spacecraft as it passes through the Earth's radiation belts, during geomagnetic storms, and particularly when immersed in low-density plasma. The amount of charge deposited on the skin of the spacecraft will vary from one place or one material to another, establishing a difference in electric potential between them. This voltage difference can provoke electrical discharges—not unlike strokes of lightning—between different areas on the spacecraft, or between the different materials that are employed for thermal control.

Free electrons in space can also induce electrical charges on cable insulation and other non-conducting materials *within* the spacecraft, leading to similar and more harmful electrical discharges there. The effects can seriously damage components and subsystems, burn out power supplies and set off automatic commands in the spacecraft control system.

Highly energetic atomic particles—and most especially trapped electrons in the radiation belts—are able to penetrate the protective shell of a spacecraft and induce *deeper* electrical charges within electronic and computer components that can cause malfunctions, serious damage and even equipment failure.

In passing through the spacecraft, its computers and other equipment, high energy particles will also ionize some of the material through which they travel, leaving a trail of electrically-charged "particle tracks" across and through circuit boards and other critical components. Intruding particles that are less energetic can induce harmful surface charges on sensitive equipment. Spacecraft control systems are among those affected by either deep-charging or space-charging, leading to the possible loss of the spacecraft itself.

☉ ☉ ☉

The thousands of spacecraft that have operated in space in the course of the last half century have provided what is now a long record of equipment malfunctions and failures, most of which are attributed to energetic electrons trapped within the magnetosphere, solar energetic particles, and cosmic rays. Such events are sufficiently common to be separated into different categories of cause or effect.

Among the classifications routinely used are electrostatic discharges (ESD) due to surface or deep dielectric charging, and several classes of single event effects (SEE) that are traceable to a single high-energy particle. Most common among the latter are single event upsets (SEU) in micro-circuits, which occur most often when a heavy ion deposits enough charge on a sensitive circuit element to cause it to change state: the equivalent of throwing a two-pole switch from OFF to ON, or ON to OFF.

Single event upsets are a common cause of disruptions and even failures in electronic equipment in space. Among the critical functions affected are the spacecraft's orbital control and stabilization system. A great many military and commercial telecommunications satellites fly in geosynchronous orbits at an altitude of about 22,200 miles above the Earth's surface, near the outer edge of the outer radiation belt, which makes all of them highly vulnerable to damage from single event upsets. The galaxy of GPS spacecraft, which operate within the outer radiation belt, are among those that have experienced many costly upsets of this kind.

Damage to Other Space Equipment

Most spacecraft are powered by solar cells, which in space—as on Earth—absorb radiative energy from the Sun in the visible and infrared regions of the spectrum and convert it into voltage, and hence available electric power. But when taken above the Earth's atmosphere, solar cells are directly exposed to a continuing barrage of high-energy atomic particles as well. These come not only from the solar wind and solar eruptions and from more distant sources as cosmic rays, but also from confined streams of highly energetic particles that circle the Earth in the two radiation belts.

High-speed atomic particles that slam into the face of solar cells are like grains of gravel thrown against the windshield of a speeding automobile, reducing their transparency and hence efficiency in converting radiant to electrical energy.

Particles from a single, moderate intensity solar event can reduce the efficiency of solar cells by about 3%, a drop sufficient to shorten the orbital lifetime of the spacecraft. CMEs and a major flare on March 13, 1989 chopped years from the designed lifetimes of more than a dozen satellites in geosynchronous orbits.

Among those affected by these events was a NOAA *GOES* weather satellite, one of two then positioned over the western and eastern halves of the continental U.S. to provide the pictures of clouds and storm systems seen daily on broadcast television. In one day these great blasts from the Sun—whose direct effects on the Earth were felt for about six *hours*—took about six *years* from the expected lifetime of the spacecraft. Two years later, in October, 1991, other large CMEs reduced the life of all three of the *GOES* spacecraft by about two years. As a result, NOAA and NASA had to accelerate the funding, building, testing and launching of replacements, years earlier than originally planned.

Incoming particles that pass through digital imaging devices, like telescopes of various kinds, can cloud and obscure the image obtained. Any charged particles can do this, including electrons and the many secondary particles that are released internally when a more energetic particle, such as a solar proton, strikes the material surrounding the detector. When a cloud of heavy ions from a solar flare or CME-driven shock wave reaches a spaceborne telescope that is pointed at the Sun, the electronic "noise" it produces can saturate the photo-sensing element. Images recorded at these times look much like the view through the windshield of an automobile driving at night in Wyoming through a blizzard of blinding snow.

Protecting Against Damage From High-Energy Particles

Before it enters the atmosphere, an accelerated electron can pass through about a quarter inch of aluminum, and a solar proton at least some six times farther, through an inch and a half or more. Thus the light-weight exterior shells of most spacecraft provide little if any shielding against either of them.

As noted earlier, for galactic cosmic rays with energies in the range of a billion electron volts (1 GeV) there is no practical brute-force method of shielding for either the spacecraft or equipment within it, given the weight of thick, denser metals like lead or steel. Thus galactic cosmic rays can shoot right through a spacecraft, from top to bottom, side to side, or end to end. Nor will they be stopped or significantly slowed by conventional materials that encase computers and other vulnerable electronic equipment.

Another limitation in using metallic shielding is that of the secondary particles that are produced when the primary high-energy particles are stopped within it. At some point, depending on the thickness of the shielding material, these secondary showers can create deleterious effects that surpass the potential damage from higher-energy primaries.

Partial engineering solutions to these problems involve the classic trade-off between insurance and its cost: in this case, between level of protection and launch weight. In practice, while it is possible to shield against less energetic particles, and to make greater use of more costly, radiation-hardened electronic components, it is prohibitively expensive to provide 100% protection of any complex spacecraft component or instrument against high-energy cosmic rays and solar protons.

Most satellites today—due to their increased complexity, greater reliance on micro-electronics, and increasing use of on-board data processing and storage—are more susceptible to the perils of space weather than those of yesteryear. The commercial use of space, in particular, calls for lower launching costs, lighter spacecraft with even less shielding, and off-the-shelf components in place of those that are radiation-hardened.

Since future spacecraft are likely to be smaller and lighter, and therefore provide less redundancy and use even more-miniaturized and complex electronic systems, they will be more vulnerable to the damaging effects of high-energy atomic particles. In addition, with more telecommunications satellites in high geosynchronous orbits and space missions of greater complexity, more spacecraft will travel into harm's way, beyond the greater safety of low Earth orbits.

With these trends we can expect more frequent single event upsets and more costly disruptions and failures in space operations.

☉   ☉   ☉

There are four ways of minimizing radiation hazards to spacecraft and space equipment. The first three—which in most cases prove costly or impractical—are more shielding, greater redundancy, and a more conservative choice of orbits and flight paths.

The fourth is earlier warnings of hazardous times and of specific solar and magnetospheric events. Periods of expected high (or low) solar activity can be foretold, as can the *probabilities* of solar flares, CMEs and geomagnetic storms. More useful predictions and alerts depend upon the constant surveillance of the Sun, from multiple spacecraft and vantage points. They also require advance "sentinels" like the *ACE* spacecraft, stationed slightly closer to the Sun than we, to sense and evaluate oncoming streams of plasma in the fast solar wind and CMEs in order to extend the warning times of those most likely to impinge upon the Earth.

Impacts on Telecommunications, GPS, and Navigation

The electromagnetic waves that today carry information of all kinds are everywhere and ever present in our lives: spreading not only around the world but in some cases far beyond it, through the heliosphere and throughout the Galaxy.

In but four hours, television programs broadcast to viewers in Chicago will have already reached the orbit of Neptune, and in less than a day have spread beyond the heliosphere to fill the wide open spaces where other stars abound. Electromagnetic waves carrying live television programs featuring Jack Benny, Red Skelton or Lucille Ball are at this time arriving at places more than fifty light-years away, having already passed five of the twelve brightest stars in the sky. Sirius, a near star that is the brightest of these, had its chance to watch these classic broadcasts, live from the U.S.A., more than forty years ago.

☉   ☉   ☉

Our use of electromagnetic waves to convey information was but a century ago limited to telegraphy; for carrying voice messages from hand-cranked phones over local telephone lines; and for the transmission of code or spoken words in wireless radio messages within a limited segment of what was then called the "short-wave" radio band.

The full extent of the radio-frequency spectrum—a range of wavelengths spanning ten orders of magnitude and divided into nine frequency bands from ultra-low (3000 cycles/second) to super-high (30 billion per second)—has now been commandeered for modern needs. To meet the requirements of myriad users, the nine frequency bands are now divided and subdivided into more than a thousand separate but closely-packed segments—such as *aeronautical radio navigation*, *TV channels 7-13*, or *radio astronomy*—each allocated by international or federal commissions for a highly specific purpose.

Today, ordinary people rely on electromagnetic waves every day of every week for cell, cordless or hard-wired telephones; to receive radio and television broadcasts; for connections between computers and to the internet; and for remotely locking the car, opening the garage door or a host of other wireless applications. Electromagnetic waves are also used for an expanding list of civil, military and commercial purposes, which range from the transfer of transactions from automated teller machines and cash registers to communications of all kinds between points on land, sea, in the air, and in space.

And most of them are vulnerable in one way or another to the changing moods of the Sun.

Direct and Indirect Reception of Radio Waves

Radio waves sent outward from a broadcast TV tower or a cell-phone held in your hand spread outward in expanding circles, like ripples on a pond: but in this case in three dimensions and at the speed of light, 186,000 miles per second. Like waves in water, their advance in any direction proceeds along straight lines from the source.

This means that transmitted radio signals are unable to follow the curvature of the Earth to continue like a bird or an airplane beyond the horizon. The limit of their direct line of travel depends upon the height of the transmitter and the lay of the land, but in most instances is considerably less than 100 miles. For a person five or six feet tall standing on a flat reach of land, the horizon is less than three miles away; and for a dachshund, it is very near at hand.

Electronic signals, including broadcast radio and TV, can be sent beyond the horizon by the use of appropriately-sited relay stations and long-distance cables. And more commonly today, by re-transmission from telecommunications satellites from whose vantage point—more than 4000 miles high—the visible horizon of the Earth extends for thousands of miles in all directions.

Signals sent outward in low frequency bands of the radio spectrum—including among others, most amateur and all AM radio transmissions—are able to make it beyond the horizon without added help by forward reflection from the ionosphere: the electrically-conducting layers of free electrons and ions in the upper atmosphere that lie far below the lofty heights at which telecommunications satellites fly. Depending on its structure and density at the time, the ionosphere can for other radio frequencies serve as a polished mirror to allow them to go over the horizon and extend their range by repeated reflections far beyond it.

☉ ☉ ☉

It was through ionospheric reflections of this kind that in 1901, the twenty-seven year-old Guglielmo Marconi sent the first wireless signal between the Old World and the New, by successfully transmitting ••• , the three dots of the letter S in Morse code from England, "round the protuberance of the earth", to a receiver in Newfoundland, more than a thousand miles away.

Today, with adequate power and the application of very low frequencies (using wavelengths more than sixty miles long!), radio signals are bounced back and forth in the space between the Earth and the lower ionosphere all the way around the planet.

Electromagnetic signals transmitted in higher frequency bands—including the segments of the radio spectrum allocated to satellite communications, GPS, FM radio, and broadcast television—are not reflected by the ionosphere and are thus able to pass through it with little attenuation, to continue outward into space. This is why un-named, faraway planets can today receive live broadcasts of *I Love Lucy* while a viewer in the 1950s, but tens of miles from a transmitter located on the other side of a hill or down in a valley, could not.

Role of the Sun and Solar Variations

All of the telecommunications signals that pass through or are reflected by the concentrated layers of electrons in the ionosphere are affected by the Sun and solar variability, since it alone creates and sustains the ionosphere. Solar flares, CMEs, and magnetic storms—the products of solar disturbances—provoke abrupt and drastic changes in the ionosphere that can weaken, distort or temporarily block signals which under other conditions would be more cleanly reflected or allowed to pass.

The daily rotation of the Earth—carrying darkened regions of the globe into and then out of the direct blast of solar short-wave radiation—forces dramatic day-to-night changes in the density and structure of the ionosphere that exert a major impact on the range of reception of radio signals.

Day-to-day changes in the level of magnetic activity on the Sun and hence in the amount of short wave radiation received at the Earth alter the density of free electrons in the ionosphere, as do year-to-year changes in the course of the 11-year solar cycle. These solar-driven variations in electron density control not only the *reflectivity* of the ionosphere but its *transparency* and *homogeneity* as well.

Satellite communications and all others which make use of satellite repeaters rely on higher radio frequencies that are able to pass *through* the ionosphere. But these too are subject to solar-driven perturbations that not only alter its transparency to waves of different frequencies but also introduce irregularities in its structure. The radio waves that pass through a rippled region of this kind are distorted, introducing scintillations in the signal received, which are not unlike the twinkling of a star.

Among the many telecommunications systems affected by changes in ionospheric transparency and homogeneity are communications between spacecraft and the ground; marine and aircraft navigation; GPS signals; guided missile systems; and all applications, including satellite telephones and network television that rely on satellite repeaters operating well above the ionosphere in geosynchronous orbits.

Nor can the effects of the Sun on telecommunications be avoided by sending signals along wires or through under-ground or under-sea cables. These too, are affected by the Sun when CMEs and flares disturb the magnetosphere, and through this link induce troublesome electric currents in long conductors of any kind at or below the surface of the Earth

Impacts on GPS and Other Navigation Systems

The *Navstar* Global Positioning System—so extensively utilized today—relies on a fleet of twenty-four GPS spacecraft, each equipped with radio receiver-transmitters and atomic clocks. These circle the Earth at an altitude of about 16,000 miles, about three-fourths as high as the geosynchronous orbits (22,200 miles) where telecommunications satellites fly. At this height the GPS spacecraft operate in an unfriendly world, immersed in swarms of energetic electrons in the middle of the Earth's outer radiation belt.

Whether on the ground, at sea, in an aircraft or on another spacecraft, a GPS receiver can determine its own position based on the differences in the elapsed times between the transmission and reception of signals from four satellites in the GPS fleet, which utilize frequencies sufficiently high to pass through the ionosphere.

The times of travel of the GPS-to-ground signals are affected by differences in density in different regions of the ionosphere through which the four transmitted signals pass. The precision of measurement is also affected by ionospheric scintillations. While endeavors are made to compensate for at least some of these effects, transient variations in ionospheric conditions can still affect the accuracy of any GPS result. Errors of this nature are encountered about 20% of the time during the years when the Sun is most active.

Because of the altitude and the region of near-Earth space in which it must operate, the electronic equipment employed on GPS spacecraft is also highly vulnerable to deep dielectric charging and other damaging effects from high-energy particles, which are thought responsible for many single-event upsets in its operation.

The venerable *Loran* (Long Range Navigation) system that for more than sixty years was maintained by the U.S. Coast Guard to serve navigators on ships at sea, was based, like GPS, on differences between the times of receipt of radio signals from two or more different radio transmitters. For *Loran* these signals came, however, from fixed stations on land, and in order to reach distant vessels far over the horizon were carried on very low radio frequencies that reach them by repeated skips between the reflective ionosphere and the surface of the sea.

It too, was vulnerable to solar and magnetic disturbances which could abruptly alter the density and reflective properties of the ionosphere and for a time thwart attempts to use the system. Following the major solar flare of March 13, 1989, for example, the system was effectively disabled for more than four hours.

Effects on Electric Power Transmission

The delivery of electric power to densely populated areas of the United States has been severely disrupted or cut-off altogether on repeated occasions in the past following large solar disturbances and major geomagnetic storms. In several instances, the total costs of the solar-induced outages amounted to tens of millions of dollars, and in one case reached the level of financial losses encountered in the course of major floods and hurricanes. It has been shown, moreover, that space weather conditions affect the wholesale market for electricity even at times of reduced solar and geomagnetic activity.

The major disruption most often cited is the power blackout in the Province of Quebec and electrically-linked regions of the northeastern U.S. that followed the exceptionally large flare, CME and geomagnetic storm of March 13, 1989. While the 1989 power failure was unusually extensive and costly, other disruptive events of the same nature and origin had been experienced before.

Solar disturbances that tripped transformer banks, damaged equipment and disrupted electric power in wide areas had affected this country and Canada in 1940, 1958, 1972, and later in 2000. In each case they followed major magnetic storms and auroral displays in maximum years of the Sun's 11-year cycle. Similar solar conditions are next expected in the period from about 2011 to 2015.

The Power Blackout of 1989

The solar eruptions that initiated the chain of events that caused the massive power failure in March 1989 came from an extremely extensive and magnetically-complex region on the surface of Sun. This immense area—more than 40,000 miles across and roughly half as wide—contained a tangle of very strong and oppositely-charged magnetic regions, including a large collection of sunspots, some of enormous size.

To solar observers who through telescopes saw this menacing collection come around the eastern limb of the Sun, or watched the sunspots drift with solar rotation toward the center of the solar disk, it must have seemed like an approaching armada of gun-laden galleons sailing in tight formation. And indeed, in their fourteen-day passage across the face of the Sun the assembled group fired off nearly two thousand flares and lobbed at least three dozen large CMEs outward toward the planets.

The impact on the Earth of the greatest of these CME eruptions was the largest magnetic storm ever recorded, accompanied by colorful displays of the aurora borealis seen far south of their expected limits in Canada and the northern U.S.: this time reaching as far south as Southern California, Arizona and Texas and on into Mexico, Central America and the islands of the Caribbean. In Europe, where they are most often seen in the skies of northern Scandinavia, aurorae appeared as far south as Spain, Portugal and Hungary.

But the geomagnetic storm that produced these awe-inspiring sights had quite another impact on the electric power industry and its customers in Canada and the northeastern U.S. There, where the brunt of the magnetic changes were felt, it generated unwanted electrical currents at ground level that found their way into transformers and power lines, disabling generators, destroying equipment, and setting off a chain of disruption and system collapse. Hardest hit and the epicenter of damage and destruction was the Hydro-Quebec Power Company in the city of Quebec. It alone suffered losses in that one event of at least ten million dollars, and its customers lost many tens of millions more.

The same storm caused power blackouts in quite separate places in the U.S., from Maryland to California and as far south as Arizona. Among the many casualties was a massive ten million dollar transformer at a nuclear power plant in New Jersey that was damaged beyond repair.

Effects of the 1989 super storm were also felt across the ocean in Sweden, where equipment and power were also disrupted. It also touched neighboring

Finland, although there, the preventive care that had been paid to the design of circuits and system elements kept them from sustaining damage or loss.

How Magnetic Storms Disrupt Power Systems

The impact of high-speed plasma clouds from the Sun on the Earth's magnetic field greatly increases the flow of electric current in both the magnetosphere and ionosphere. And although sixty miles and more above the surface of the Earth, variations in these upper atmospheric current systems provoke rapid rates of change in the surface magnetic field of the planet, leading, in turn, to marked differences in the electric potential at different locations on the surface of the ground.

The electric potential of the Earth's surface at places ten miles apart can differ by as much as 160 volts. In time these inequalities will be gradually drained away by a leakage of sub-surface electric currents passing through the weakly-conducting ground. But when grounded wires, sub-surface cables, pipelines, iron railroad tracks or other metallic conductors happen to connect regions of different surface potential together a short circuit ensues, much as when a metallic object is inadvertently placed across the poles of a storage battery. The sudden direct currents that instantly flow through these shunts of opportunity—typically 10s to 100s of amperes—are known as geomagnetically-induced currents or GICs.

When a direct current of this kind, passing (in the wrong direction!) *up* a ground wire reaches the windings of a high voltage transformer at generating plants or substations it initiates a chain of disruption. This includes transformer saturation, overheating and other damaging effects that can ultimately trigger protective relays throughout the system to which it is connected.

In extreme cases the effect can ripple through an entire electric power grid, leading to a possible collapse of the network, pervasive power blackouts, and permanent damage to the immense transformers and other large and seldom-ordered components, like large transformers, that are both very costly and very slow to replace since they are individually manufactured on demand.

Where Solar-Driven Power Outages Most Often Occur

Damaging GICs occur predominantly in higher-latitude regions where the most violent magnetic disturbances occur and aurorae are most frequently seen. In the U.S. they are most prevalent along the top two tiers of the connected forty-eight states. In addition within the upper tier, the eastern half of the U.S. is more likely to be hard hit by the effects of GICs than states west of the Mississippi. This inequity stems from geological differences in the conductivity of the subsoil in the two regions, which is more rocky in the Northeast and hence more resistant to the flow of equalizing sub-surface electric currents; and also from differences in population density.

North America, moreover, is more prone to damaging GICs than is northern Europe or Asia, because of the present, western offset of the north magnetic pole, that allows more incoming charged particles into the top of the western world than at equivalent geographic latitudes in northern Europe and Asia.

Thus it is not by chance that so many of the severe solar-induced power disruptions in the last sixty years have struck the eastern provinces of Canada, and in this country, large metropolitan areas in the Northeast and North central states. This is not to say that other regions are by geography immune. The severe magnetic storm of August 4, 1972, for example, had direct impacts in British Columbia and down into the U.S. Midwest.

Today the major determinant of where power blackouts will occur in our country is not so much the present skewed location of the north magnetic pole or the resistivity of the soil in rocky New England, but the extent of interconnected electric power systems. With these much expanded grids—which now connect high-prone Northeastern power plants and distribution systems all the way to the Gulf States—no city is an island.

A magnetic storm as intense as the one that caused so much damage in the Hydro-Quebec power system in 1989 would today affect a far larger area. In this not-unlikely event, two vast sections of the U.S. would likely suffer a total power system collapse, requiring days to bring the entire system back on line.

The first and largest of these would likely encompass most of the eastern states, in an area reaching from Maine to northern Florida and extending westward as far as St. Louis, Memphis and Birmingham. The second would include all of Washington, Oregon and Idaho and parts of far-northern California and Nevada. The probable cost to power companies alone of this one super event has been estimated to be at least five billion dollars: a figure which would be dwarfed by the total economic losses in the communities involved and throughout the nation.

SOME MAJOR ELECTRIC POWER DISRUPTIONS IN THE U.S. AND CANADA TRIGGERED BY SEVERE GEOMAGNETIC STORMS

STORM DATE	AFFECTED AREA
1940 March 24	Eastern Canada, Central and Northeastern U.S.
1958 February 10-11	Eastern Canada, U.S. Upper Midwest
1972 August 4	Western Canada, U.S. Upper Midwest
1989 March 13	Southern Canada, Northern U.S.
2000 July 14	North America
2003 October 29-30	Northeastern U.S.

Effects of Geomagnetically-induced Currents on the Cost of Electricity

Geomagnetically-induced currents provoked by solar activity and geomagnetic storms can affect the price of electricity, regardless of the magnitude of the disturbance, through a reduction in the ability of electric power distribution systems to deliver low-cost power to areas of higher demand.

A recent study of a major Northeastern power grid serving almost 10% of the population of this country found that the cost of GIC losses to consumers amounted to about half a billion dollars in the 18-month period that was examined. The price impact was not surprisingly a strong function of the severity of the geomagnetic storm, but less intense storms also took their toll. Following a major storm like that triggered by the Bastille Day flare and CMEs of July 14, 2000 the wholesale price of electricity was more than doubled, and the average over all geomagnetic disturbances, large or small, was a not-insignificant increase of 3.3%.

Early Signs of Solar Interference in Communications

Pipelines and cables—whether on, above, or beneath the surface, or laid across the ocean floor—can also be affected by geomagnetically-induced currents. These too, are electrical conductors in a position to provide a short-circuit path between separated surface areas of different electric potential.

In 1926, at a time when far more was known about sunspots than about geomagnetically-induced currents, Guglielmo Marconi—then fifty-two and still much involved in telecommunications—called attention to the fact that the times when undersea cables and land-lines were thrown out of action seemed always to coincide with the appearance of large sunspots and intense aurorae. And that these were also times of frequent fading of high-frequency wireless radio transmissions.

The occasional presence of "anomalous currents" in telegraph wires had been commented upon as early as 1847, although at that time neither geomagnetically-induced currents nor their solar and magnetospheric causes were known or understood. Nor was it probably noted that 1847 marked the peak of the 11-year solar cycle (cycle #9) that ran between minima in 1843 and 1856. In fact, at that early date very few people anywhere were aware of the Sun's cyclic behavior, even though Heinrich Schwabe had published his landmark paper announcing that finding four years earlier, in 1843 in the scientific journal *Astronomische Nachrichten*.

But few had read it at the time and fewer still were prepared to accept what he claimed as fact. Schwabe's Cinderella paper would have to wait four years more, until 1852, when Doktor Professor Alexander von Humboldt— not an amateur like Schwabe, but a celebrated professional scientist of world renown—endorsed the pharmacist's discovery by calling attention to it in his four-volume series of widely-read popular books, entitled *Cosmos: A Description of the Universe*. Then as now, for Schwabe (whose life was suddenly changed) it was not so much what you knew, but whom you knew.

Some Effects of GICS on Telecommunications Cables

Following the severe magnetic storm of February 11, 1958, telephone and teletype signals carried in the first-ever transatlantic telecommunications cable—running between the shores of Newfoundland and Scotland—were disrupted for almost three hours due to the excessive geomagnetically-induced currents that were imposed. Telegraphic messages were garbled, and words politely spoken at one end of the line were heard at the other as squeaks and whistles: like sounds from a flock of starlings.

Geomagnetically-induced currents caused by the great solar flares of early August, 1972—and by unseen CMEs which at that time had yet to be discovered—succeeded in disabling an AT&T telecommunications cable that ran from Chicago to the west coast. One of the effects was to cut off all long-distance telephone traffic between Chicago and Nebraska.

Since 1990, four out of five transoceanic telephone calls are carried in cables beneath the ocean. During the record-breaking magnetic storm of March 14, 1989, the first-laid transatlantic voice cable running along the ocean floor was rendered almost inoperable when large electric currents were induced in it by the storm-induced difference in electrical potential between the cable's terminal stations in New Jersey and England. Nor has the switch to wider-bandwidth optical fiber cabling made much of difference in this regard, since the bundles of glass fibers are bound with conducting wire. These too, pick up and transmit unwanted electrical currents at times of severe magnetic storms.

Damage to Pipelines

Geomagnetically-induced currents can also accelerate corrosion in metal pipelines, such as those used to carry oil or natural gas over distances of hundreds to thousands of miles. The intrusion of magnetically-induced currents can also interfere with the technical systems that are employed in pipelines to combat corrosion.

Particularly susceptible to induced pipeline corrosion are any diversions or irregularities: including places of departure from straight-line runs—such as bends and branch points—and joints where different metals meet.

Impacts of Geomagnetic Storms on Geological Surveys and Explorations

The onset of a geomagnetic storm, although playing out a thousand miles and more above our heads, is immediately registered as an abrupt change in the magnetic field at the surface of the Earth. There it is readily apparent as erratic behavior in compass needles, and readily captured for later study by continually-running magnetometers at magnetic observatories around the world.

We usually think of magnetic compasses as hand-held direction finders used by Boy Scouts and other hikers, on collapsible tripods by road-side surveyors, or freely supported in brass binnacles bolted to the deck on the bridges of ships at sea. Magnetic storms can perturb all of these familiar uses and in addition—often at great cost—the more sophisticated and automated applications in which the Earth's magnetic field is relied upon as a fixed directional reference.

One of these applications is in guiding the downward course of rotating bits used in drilling oil wells. Here any perturbation in the reference direction results in an immediate change in the direction of travel of the drill bit, which can lead to costly errors and equipment damage. Thus to drillers and surveyors a major magnetic storm is no less and probably more of a threat than the sudden onset of severe weather.

This was the case for a number of North Sea oil companies when the severe magnetic storms that followed the very large CMEs and solar flares of early March, 1989 forced them to abandon all efforts to drill.

The same event displaced compass needles for a time by a whopping 10° from the direction of magnetic north, with effects on navigators and surveyors around the world.

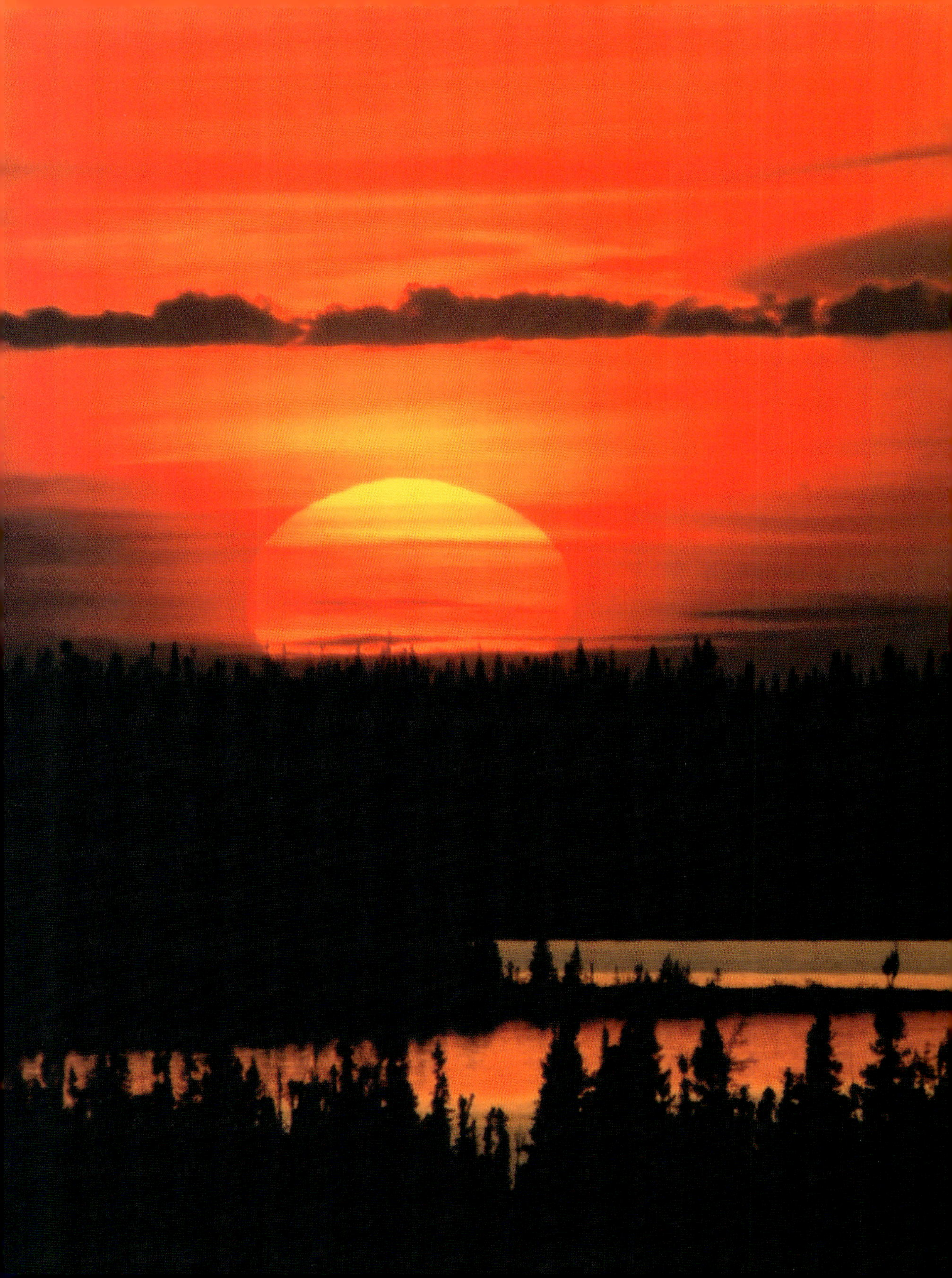

EFFECTS OF THE SUN ON WEATHER AND CLIMATE

A Brief History

It has long been known that the Sun provides almost all of the energy that powers the weather machine, including the mighty forces that drive winds and storms, push ocean currents, and cycle the water between the surface and the air. What is not as well established—even today—is the extent to which variations in the weather and climate arise from fluctuations, short or long, in the energy the Sun delivers at our door.

This practical question, so often asked, has been around in one form or another for a long, long time. For at least 200 years natural philosophers and astronomers, then solar physicists and meteorologists, and now climatologists and paleoclimatologists and oceanographers have tried to find the answers, in the hope of achieving practical weather and climate prediction.

Concerns about the Sun's constancy are common in early religions, and probably as old as human thought. Almost as old, we may presume, are the intuitive feelings that since the Sun is the obvious source of heat and light and day and night, might it not also control the seemingly random vacillations of weather and longer-lasting changes in climate?

Soon after Galileo and his co-discoverers first saw dark spots on the face of the Sun, in 1609, it was apparent that the number of sunspots (and their sizes and positions) changed from day to day. That these variations on the surface of the Sun might possibly affect weather in Europe or elsewhere on the Earth must have been to them an obvious deduction. And one that has never gone away. The conjecture that sunspots might somehow affect the weather was common enough that about a hundred years later, Jonathan Swift in 1726 wove this common presumption into his tales of *Gulliver's Travels*.

There he tells of the floating island of Laputa: a mythical land populated by philosophers and astronomers who—equipped with telescopes better than ours—were not only obsessed with the sky, but always troubled by what they saw. Among their many apprehensions was a fear that the face of the Sun might become so covered with dark spots that it would no longer provide sufficient heat and light to the world. When the Laputians met an acquaintance early in the day, he tells us, their first question was not the customary "How are you?" but "How did the *Sun* look this morning?"

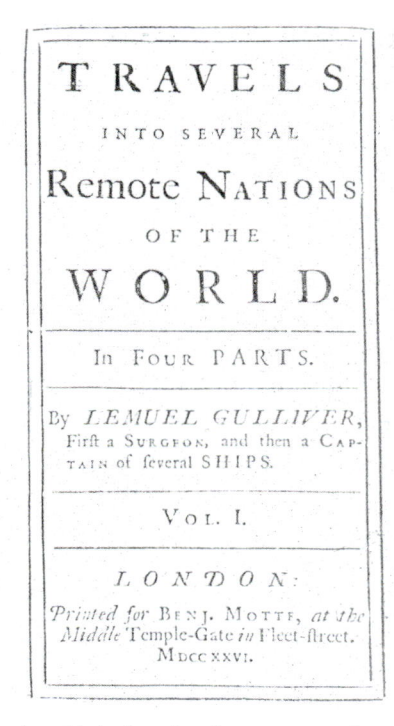

Jonathan Swift (1667-1745) and the title page of his book published under the pen name Lemuel Gulliver in 1726. Throughout Swift's life, astronomers and other learned people were well aware of an unusual paucity of spots on the face of the Sun persisting for fully 70 years, from about 1645 until 1715: a paradox which probably influenced Swift to include a fanciful but related episode in Gulliver's Travels.

Real astronomers in the eighteenth and nineteenth centuries who looked at the real Sun in the real world were also intrigued by the possibility of a Sun-weather connection. Quite apparent to any of them was the enormous practical benefit to society—in an age when the prediction of weather was based almost entirely on accumulated lore and seasonal expectations—were a clear connection to be found linking the presence or absence of easily observed features on the disk of the Sun to local or regional weather conditions.

☉   ☉   ☉

Many claimed to have found it, including Sir William Herschel, the celebrated astronomer and builder of large telescopes, whose paper published in 1801 in the *Philosophical Transactions of the Royal Society* reported his own discovery of a persistent relationship between the prevalence of sunspots and the price of a bushel of wheat on the London market. Based on his own and earlier observations of the Sun since about 1650, he found that during protracted periods when sunspots were scarce the price of wheat was always higher. Herschel reasoned that fewer spots on the Sun denoted abnormality and an

Sir William Herschel (1738–1822), widely considered the leading astronomer of his day, was among those who suspected a strong link between protracted periods of unusually low or high solar activity, measured in the numbers of sunspots seen, and the Earth's weather. His conclusions, published in 1801, preceded by almost half a century the realization that sunspots followed a short-term cycle of about 11 years.

accompanying deficiency in the amount of heat the Sun released, leading to poorer growing conditions, diminished agricultural production, and through the inexorable law of supply and demand, higher commodity prices.

When Herschel published his paper, the cyclic 11-year rise and fall in the number of sunspots was not yet known, and wouldn't be for nearly half a century. The belated discovery of this strongly periodic feature in the annually-averaged numbers of spots on the visible surface of the Sun was announced, as we have noted earlier, by Heinrich Schwabe in Germany in 1843.

In time Schwabe's discovery proved to be a seminal revelation into the physical nature of solar activity and variability. In the mid-nineteenth century, however, the principal effect on scientists and many other people was a rush to find statistical evidence of meaningful connections with other phenomena—and particularly the weather, in the hope of finding keys to practical weather prediction. Schwabe's announcement triggered an avalanche of claims that continued unabated for years, each purporting to have found meaningful correlations between the sunspot cycle and a host of things meteorological, hydrologic, oceanographic, physiological, behavioral, and economic.

Sir Norman Lockyer, the Victorian solar physicist who founded and for fifty years edited the journal *Nature* was an early champion of these searches, counseling in 1873 that "…in Meteorology, as in Astronomy, the thing to hunt down is a cycle." He had found one, himself, the previous year, in records of the intensity of monsoon rains in Ceylon that were clearly linked, he said, to the ups and downs of the 11-year sunspot cycle; adding, with characteristic modesty, that based on his discovery, "…the riddle of the probable times of occurrence of Indian Famines has now been read."

Indeed, until relatively late in the 20th century almost the only tool available for the investigation of possible influences of solar variability on weather and climate was the statistical comparison of indices of solar activity with contemporary meteorological records. And while some of these searches proved valuable as probes and tests of a complex system, most of what was found seemed soon to go away, and few of the correlations that were claimed stood up to rigorous statistical tests. Without the buttress of a solid physical explanation, none proved to be of any significant value in practical weather or climate prediction. The required leap of faith between what was seen on the Sun and what was felt at the bottom of an ocean of air on a small planet 93 million miles away was simply too great, in the absence of a fuller knowledge of what happens in between.

Beginning in the 1870s, heroic efforts were made by the American astronomer Samuel Pierpont Langley and others to put the question on a more solid basis by attempting direct measurements of the Sun's radiation from the top of Mt. Whitney and Pike's Peak and other high-altitude stations. Through the persistence of Langley's assistant and successor, Charles Greeley Abbot, attempts to identify possible changes in solar irradiance from the surface of the Earth were continued at different stations around the world through the first half of the ensuing century: but to little avail, due to difficulties in absolute calibration and larger uncertainties in allowing for the variable absorption and scattering of solar radiation in the intervening atmosphere.

In spite of more than a century of dedicated effort, the very practical question of whether and how solar variability affects the Earth's weather and climate would remain largely unanswered until relatively recently when needed facts were at last obtained.

The Missing Pieces

Missing and badly needed in early attempts to find answers to the Sun-Climate question were (1) a fuller understanding of the effects of sunspots and other apparent changes on the energy released from the Sun; (2) a more complete understanding of other forces that perturb the climate; (3) the availability of analytical tools to test and evaluate proposed responses to solar fluctuations in a system that was both complex and interactive; (4) a reliable record of global climatic changes, both past and present; and perhaps most important; (5) quantitative measurements of any changes in the amount of solar energy the Earth receives.

While some facts were known about the surface of the Sun two hundred years ago when Sir William Herschel proposed a causal connection between sunspots and the price of wheat, next to nothing was known about what we now call *climate*, as opposed to short-term *weather*. The same was true almost a century later when Sir Norman Lockyer laid his shaky claim to having unlocked the secrets of the Indian monsoon.

The fundamental questions were these: Through what limits *does* the energy received from the Sun vary, from day to day or year to year? *How much* does the climate vary at any place, or over the Earth as a whole? How does the climate system work, and *how important* a player is the Sun?

Today we have most of the answers, due largely to our ability to observe the Sun and global phenomena from the vantage point of space, and the modern capability of creating and testing realistic models of the entire climate system that incorporate variable inputs from the Sun. Major advances have also been made in the last few decades about the interactive climate system and the history of climate change, driven in large part by world concerns regarding global warming and its likely economic, societal and environmental effects.

Metering the Energy the Earth Receives

Many of the advances of the past few decades in understanding the effects of the Sun on climate come from direct measurements from space of the fluctuations in solar energy received at the top of the Earth's atmosphere: truly, where the rubber meets the road. These seminal measurements of total solar irradiance—initiated in 1978 and continuing today—give needed substance to modern investigations of the Sun and Climate, while providing answers to the oldest of all solar questions: How constant is the Sun?

We now know that the total solar irradiance varies from minute-to-minute, reflecting activity-driven changes on the face of the Sun; from day-to-day in step with solar rotation and the evolution of solar active regions, with day-to-day amplitudes of up to about 0.3%; and more important in terms of climate, from year-to-year with a peak-to-peak amplitude of about 0.1%, in phase with Schwabe's 11-year sunspot cycle. In years when the Sun is more active, and more spots are seen, more radiative energy is delivered to the Earth—just as Herschel had surmised, 200 years ago.

The impact of variations of this amplitude on surface temperature depends on the persistence of an increase or decrease, and the sensitivity of the climate system to solar forcing. In theory, and were nothing else at work, a sustained

Total Solar Irradiance

Irradiance Variability Components

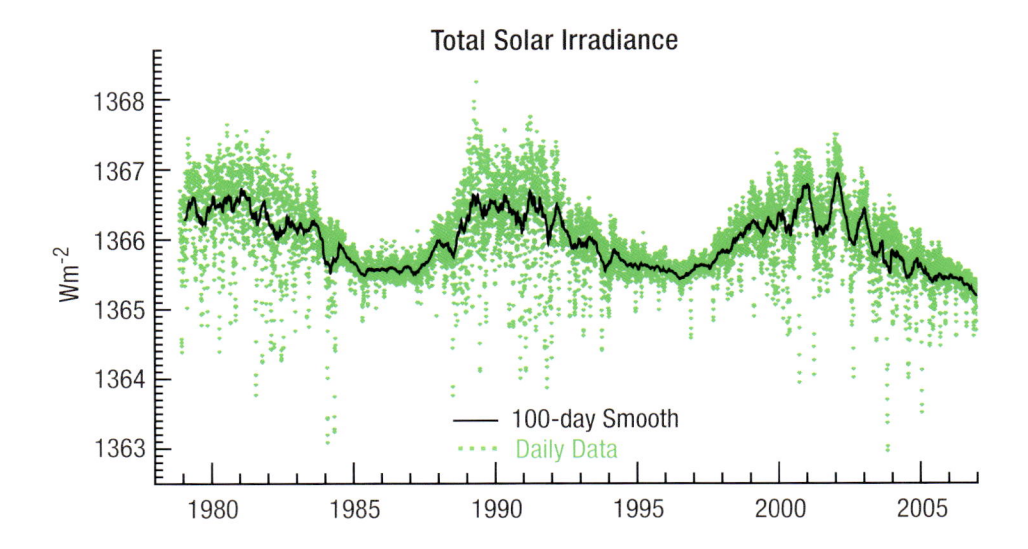

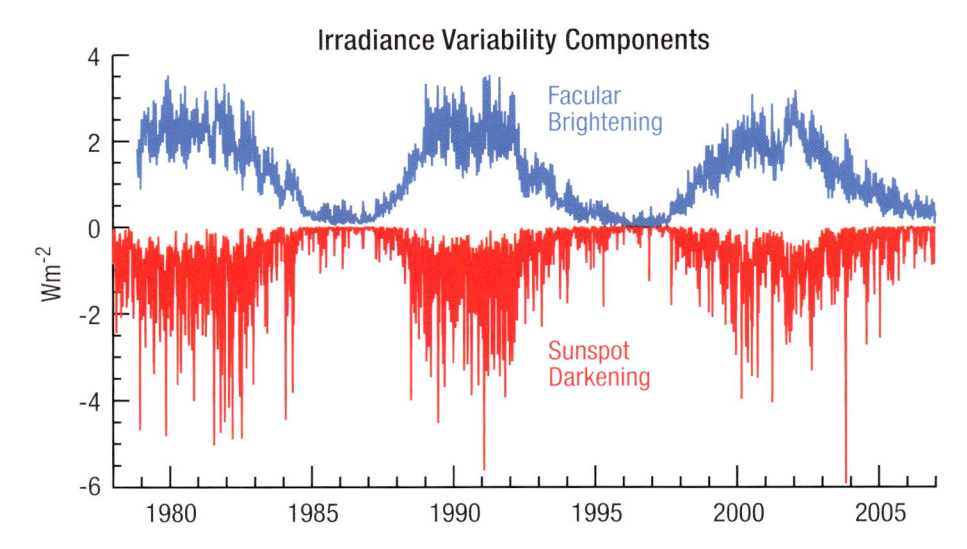

Upper: Changes in the total radiation received from the Sun at the top of the Earth's atmosphere through three 11-year solar cycles, starting in 1978 when continuous radiometric measurements of this fundamental parameter were begun. Maxima and minima are in phase with coincident changes in solar activity, with maximum radiation received in years of maximum solar activity.

Lower: Day-to-day changes in total radiation expected from daily measurements of the areas of bright faculae (which increase the radiation received) and sunspots (which act in the opposite direction) observed on the surface of the Sun. Although sunspots are more easily seen, faculae and bright regions exert a greater effect on the annual average of solar radiation received at the Earth, as is apparent in the upper figure.

increase of 0.1% in total solar irradiance can be expected to warm the surface temperature of the Earth by about 0.1°F.

And indeed, changes of this amount have since been found in averaged regional and global weather records, extending back for fifty years or more, in both the temperature of the air and in surface and subsurface temperatures of the oceans. These could easily be attributed to changes in total solar irradiance, since the amplitudes and phase of the temperature anomalies are consistent with the observed 11-year variation in the radiative output of the Sun.

But the remarkably close agreement with the ocean data, in particular, is in some ways enigmatic, for the thermal inertia of the oceans should more heavily damp the climate system's response to "rapid" fluctuations in solar energy.

It takes about three years for the upper, mixed layer of the ocean to fully respond to heat added at the surface. This means that unlike the air, its temperature at any time tells not so much about *today* as about the *past*, and more precisely of what it remembers of conditions during the last several *years*. Yet the Sun's 11-year cyclic forcing persists for only a few years in one direction before reversing itself: three years to rise from minimum to maximum, two or three years there, then a slower fall of five or six to the next minimum.

The effect of the oceans' three-year memory in responding to short-duration changes of about the same length should significantly reduce its theoretical sensitivity to imposed solar forcing. Yet the 11-year variation found in ocean temperature data is not less but *more* than what is expected.

What seems likely is that other climatic processes are indeed at work, including possible feedbacks that tend to amplify the response of the Earth's climate system to subtle solar forcing. In addition, 11-year variations larger in magnitude than in the total solar irradiance are found in the Sun's *spectral* irradiance and in its output of high-energy particles. Either or both of these could well prove responsible for amplifying the apparent response of the atmosphere to 11-year solar forcing.

☉ ☉ ☉

Other important questions still remain. Is the Sun so massive and imperturbable that the more superficial 11-year modulation of total and spectral irradiance and solar particles are the *only* significant variation in the energy it releases? Might there be other, longer-term changes? Our only continuous record of solar energy received at the Earth—from 1978 to the present day—covers but the blink of an eye in the life of the Sun. Nor has it yet allowed us to detect the effects of known fluctuations in solar activity that persist for decades and longer.

Best known among these longer-term changes are the recurrent 50 to 100-year episodes of severely suppressed activity, like the Maunder Minimum of 1645-1715, that are prominent features in both historic records of sunspots and in the far longer proxy records of solar activity obtained from dated ice cores and tree-rings. Nor did the space-borne measurements begin early enough to catch the systematic rise in the overall level of solar activity which was the Sun's dominant characteristic through the first half of the 20th century.

RECENT PROLONGED EPISODES OF LOW AND HIGH SOLAR ACTIVITY

EPISODE	DURATION
Spörer Minimum	1540 - 1450
Maunder Minimum	1645 - 1715
Dalton Minimum	1790 - 1830
Modern Minimum	1875 - 1915
Modern Maximum	1940 - Present

There could be other, deeper-lying causes of change in solar irradiance—periodic or aperiodic, related or unrelated to solar activity—that operate more slowly and on longer scales of time. And because of the thermal inertia of both the land and the oceans, we should expect the climate system to be more responsive to slower and more persistent forcing than to daily, annual, or 11-year fluctuations.

One reason for *suspecting* longer-term changes of larger-amplitude in solar radiation include an apparent correlation between the Spörer and Maunder minima of solar activity in the 15th through the early 18th centuries and particularly cold epochs of the contemporaneous Little Ice Age, and the close correspondence between features in the paleoclimate record of the Holocene epoch with what is known of the behavior of solar activity from proxy data during the same periods of time. But these apparent associations are with indices of solar *activity* and not necessarily solar *irradiance*. While irradiance changes are arguably the most likely solar cause, they are but one of several activity-related variations in the output of the star.

☉ ☉ ☉

An auxiliary method for studying possible long term changes in the radiative output of the Sun is by adding statistical evidence from dedicated, ongoing observations of the spectra and brightness of Sun-like stars. Some early studies seemed to indicate likely similarities. But these have not been substantiated and as yet the radiometric findings from other stars are far from conclusive on this matter.

Recovering the Past History of the Sun

We know now that sunspots *inhibit* the upward convective transport of energy to the photosphere and *diminish* the emergent radiation in proportion to their total projected area as seen from the Earth. Bright areas that surround sunspots and the bright boundaries of the tops of convective solar granules act in the opposite direction to *increase* the Sun's total energy output. As a result of this ongoing tug-of-war the amount of radiation the Sun emits and we receive is continually driven up and down: at times increasing the energy we receive and at others reducing it.

In the long run, when averaged over a year or so, the brighter elements prevail, as is apparent in the observed rise (of about 0.1%) in measured total irradiance in years of greater solar activity, and the ensuing fall as solar activity declines.

These clear associations linking easily-observed solar *features* to solar *irradiance*—a parameter which is far more difficult to measure and monitor—provide a kind of Rosetta Stone, or template, for translating earlier historical observations and records of the numbers of spots seen on the Sun into histories of associated changes in total or spectral irradiance.

⊙   ⊙   ⊙

Drawings and other accounts of the number of spots that were seen on the Sun on any day are available through most of the long 400-year history of telescopic solar observations. But the period in which these can most reliably be employed to recover associated changes in solar irradiation does not begin until the second half of the 19th century: on the heels of Heinrich Schwabe's discovery in Dessau.

It was then, in response to fast-growing interest in the possible terrestrial effects of the sunspot cycle, that a daily photographic patrol of the white-light solar disk was first begun: initially in 1858, at London's Kew Observatory; not long after, on the banks of the Thames at the Royal Greenwich Observatory; and in due course at other stations in other countries around the world. A corresponding set of continuous photographic images of the solar chromosphere is available from at least 1905 onward. For the period of time between about 1880 until 1978, when direct measurements began, the changes in total solar irradiance that are related to the 11-year solar cycle can be reconstructed with some confidence. What they cannot tell, however, is whether there were other, slower and possibly larger changes as well.

Nevertheless, estimates of solar radiation derived in this way from historical sunspot records resemble quite closely the long-term trends and excursions in the mean global surface temperature of the Earth for the same period. Moreover, when postulated changes of greater amplitude than those measured in the past quarter century of record are included—keyed to known, long-term changes in the overall level of solar activity—the fit with the surface temperature record of the last 100 years is remarkably good.

Based on the presumption that these long-term changes in the overall level of solar activity tell of changes of larger amplitude in solar irradiance, about half of the documented rise in the surface temperature of the Earth in the period from about 1900 to 1940 might possibly be attributed to the Sun. In the remaining years of the century just ended, the fraction of the total change in mean surface temperature that can be attributed with certainty to solar variability drops to less than 5% of the concurrent steeper rise in temperature, with the remainder widely attributed to the direct effect of ever-increasing levels of greenhouse gases. But it must be emphasized that the larger-amplitude, slower changes in solar irradiance on which these conclusions are drawn are largely speculation. The more conservative assessment which dismisses the possibility of as-yet undetected irradiance changes of longer term, is that the effect of solar changes are no motre than a ripple on the back of a gigantic swell driven by increasing greenhouse gases.

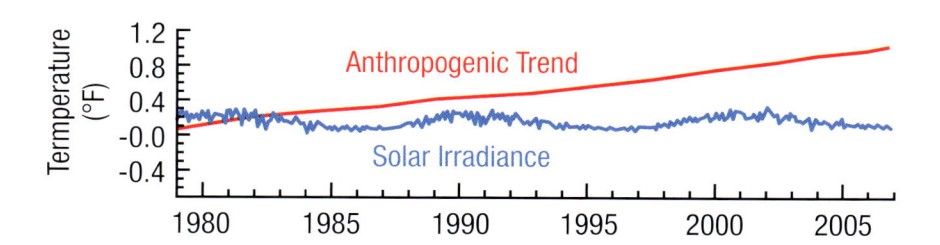

A comparison of the effects of two forcing factors on the mean surface temperature of the Earth through the period from 1978 and 2007: in blue, from recorded measurements of the total radiation received from the Sun; and in red, the calculated long-term trend due to the recorded increase in atmospheric greenhouse gases. The human-induced (anthropogenic) trend far exceeds the amplitude of the expected effect of the Sun on the Earth's surface temperature, washing out the periodic lower-amplitude effect of recorded changes in solar forcing.

How much farther back in time can we hope to improve our knowledge of past activity-related changes in solar irradiance?

Drawings of the disk of the Sun, and written records of sunspots and bright faculae are indeed available for the last several hundred years. And in almost all of these the Sun's 11-year cycle, including changes in its amplitude, are readily found. But the reliability of the pre-1850 records begins to fray when there are fewer than 365 sampled dates in any year, since the face of the Sun can vary a great deal from one day to the next, in both active and quiet years.

By this criterion the quality of the historical record of annually-averaged sunspot activity (and by association, of reconstructed changes in solar irradiance) degrades from *excellent* in the period since 1850, to *fair* from that date to about 1818, and successively *poorer* before that time.

Nonetheless, the patched-together accounts of the number of sunspots seen on the Sun—beginning in the early 17th century and continuing to this day--comprise what is probably one of the longest continuous diaries of anything in science. And this long record can now be read, both as proof of an ongoing 11-year solar cycle and as evidence of slower variations of longer term in the behavior of the Sun.

The less direct but far longer proxy records of past solar behavior obtained from the carbon and beryllium isotopes ^{14}C and ^{10}Be, described later in this section, also speak eloquently and emphatically of recurrent long-term changes in solar activity. In the most recent five hundred years of these proxy data the same long-term variations appear that are found in early historical records of sunspots and aurorae.

Effects of Solar Spectral Radiation

As we have seen, the bulk of the radiation from the Sun, in the visible and near-infrared regions of the spectrum, comes from the 10,000° F surface of the photosphere, while solar radiation in both shorter or longer wavelengths originates in higher and much hotter levels in the solar atmosphere: in the chromosphere, transition zone and corona.

In these magnetically-dominated levels in the Sun's atmosphere, solar activity plays a greater role, and for this reason the radiation they emit varies through a much wider range than does the Sun's light and heat in the visible and near-infrared.

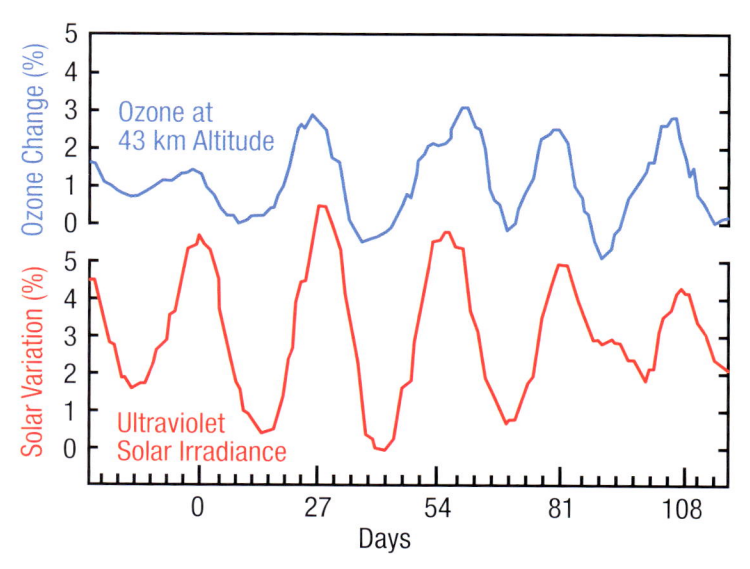

Measured changes in solar ultraviolet radiation received at the top of the atmosphere (in red) over a period of about five months, showing modulation due to the effect of the Sun's 27-day rotation. In blue are contemporaneous measurements of percentage change in the amount of ozone in the high stratosphere where ozone is created by the action of this form of solar radiation on molecular ozone.

Of particular interest are the regions of the more variable parts of the ultraviolet spectrum that control the chemical composition and photochemistry of the middle atmosphere, including the stratospheric ozone layer.

The connection between 11-year cyclic variations in solar ultraviolet radiation and the amount of ozone in the stratosphere is well established, although a quantitative relationship, based upon measurements taken during the last two solar cycles, was made more difficult by the chance occurrence of two major volcanic eruptions—which can affect the transparency of the air and the production of ozone—during the same period of time: *El Chichón* in Mexico in 1982 and *Mt. Pinatubo* in the Philippines in 1991.

Stratospheric ozone can also affect the dynamics of the atmosphere, including the circulation in the troposphere. The chain of physical and chemical processes that link solar ultraviolet radiation to stratospheric ozone and then to radiative or dynamical coupling between the stratosphere and troposphere is one of the two most likely mechanisms—on the basis of available energy—to explain apparent connections between solar activity and climate. The other is direct solar heating through changes in the total solar irradiance. The effects of either one could be either damped or amplified by possible feedback mechanisms within the internal climate system.

Past changes in the Sun's ultraviolet spectral irradiance can also be reconstructed from photographic images of the solar disk made in the visible spectrum, taken in the restricted light of the center of strong absorption lines in the solar spectrum that originate at the same higher level in the solar atmosphere as does much of the ultraviolet radiation.

Measurements of the radiation the Earth receives in different parts of the solar spectrum—from the far ultraviolet through the near infrared—have been recorded continuously since 2003 when NASA's *SORCE (Solar Radiation and Climate Experiment)* spacecraft was launched. Measurements from *SORCE*, which also carries a total radiation monitor, make it possible for the first time to examine the relative contributions of different spectral components to the total radiation received: a powerful aid in determining their possible roles as agents of weather and climate change.

Sensitivity of Climate to Solar Fluctuations

Through how many degrees should we expect the mean surface temperature of the Earth to rise (or fall) were the Sun's total output of energy to increase (or decrease) by a given amount, say 1%? A quantitative answer to this simple but important question defines what is called the climate sensitivity, which is among the fundamental parameters required in sophisticated numerical models of the climate system.

It is possible to calculate, based on fundamental physical principles, what the temperature response should be. The answer is that in the absence of any internal amplification or feedbacks, each increase of 1% in the total solar radiation received at the Earth should raise the mean surface temperature of the planet by about 1° F—or as noted earlier, 0.1° F for each 0.1% increase.

But it is more to the point to determine the sensitivity of climate to solar forcing *empirically*: that is, based on real measurements under real conditions in the real atmosphere. And this also has been done.

⊙   ⊙   ⊙

One of the most recent and extensive determinations of climate sensitivity was based on the analysis of globally-complete meteorological data sets for the twenty-five year period for which direct measurements of total solar irradiance were then available. When the known effects of the other dominant forcing mechanisms are subtracted—including El Niño/La Niña, volcanic eruptions, and the documented increases in greenhouse gases and atmospheric aerosols—a clear 11-year modulation remains. Moreover, it is in phase—as we should expect—with the solar activity cycle, in the sense that years when the planet is hotter correspond to years when the Sun is more active.

If we assume that variations in solar *irradiance* are the cause, the amplitude of the apparently solar-driven modulation in surface temperature is about two times greater, as noted earlier, than simple theory would predict: or about 0.2° F for the observed 0.1% change in total solar irradiance.

The difference of a factor of two is in agreement with a similar study of *ocean* temperature data sets, and also with what is implied in another even more extensive study of a remarkably robust correspondence between reconstructed temperatures for the last 11,000 years and proxy records of changes in solar activity during the same period. A likely explanation of what seems to be a consistent difference is a feedback in the climate-system that amplifies the direct impacts of rather small changes in solar irradiance.

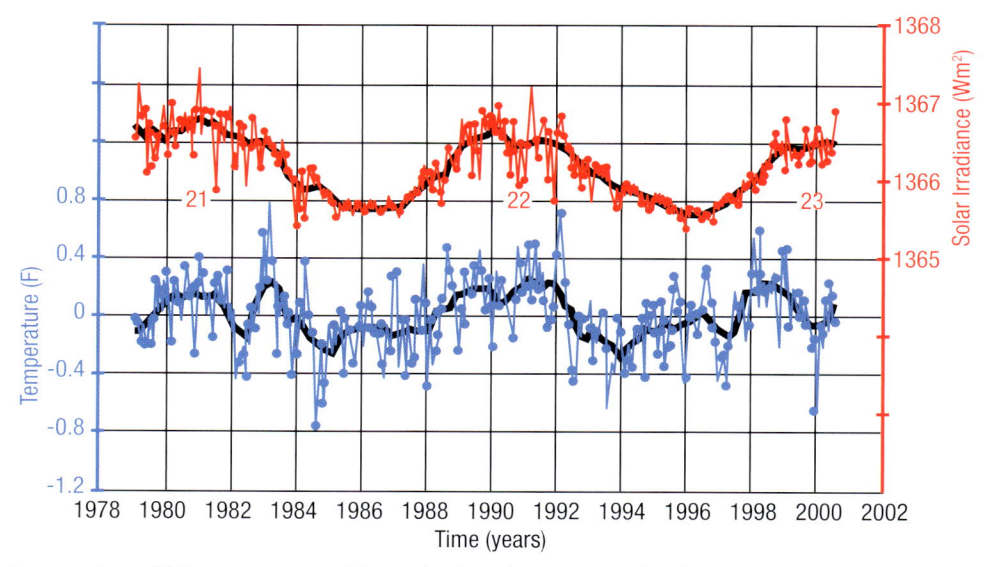

A comparison of (1) measurements of the total solar radiation received at the Earth during solar cycles 21, 22, and 23 (in red); an (2), in blue, spacecraft measurements of the northern hemisphere temperature of the lower troposphere in the same period. The effects of volcanoes, El Niño/La Niña events, and a fitted linear trend have been removed from the temperature record.

A similar investigation, based on global surface temperatures covering the entire 20th century came to a similar conclusion. In common with other modeled studies of the same kind, it was found that combining solar *and* anthropogenic (human-driven) forcing in approximately equal amounts provided a good fit to the global documented global warming of the Earth in the first half of the century. However, to fit the continued and steeper rise in mean surface temperature in the period following 1940 required additional non-solar heating, which was attributed to the rapid increase in greenhouse gases and atmospheric aerosols at this time. In this more recent period the ratio of anthropogenic to solar forcing needed to explain the observed temperature rise was about 4:1. Which infers that increased solar warming, acting alone, might explain only about 20% of the rise.

An apparent amplification of the effects of changes in solar irradiance was also found in climate models when more CO_2 and other greenhouse gases were introduced into the atmosphere.

These and other results suggest that the effect of solar irradiance variability on climate depends to some degree on pre-existing conditions and what else is at work at the time, and where, in the coupled (land-ocean-atmosphere) climate system. If so, it may help explain the will o' the wisp nature of so many of the here-today, gone-tomorrow correlations that have been claimed through the years linking weather and climate parameters with the sunspot cycle—including, among others, Norman Lockyer's purported discovery of an 11-year monsoon cycle in British India in 1872.

11-Year Solar Forcing

Beginning in the high stratosphere, some thirty miles above the surface of the Earth, any changes in the atmosphere are directly driven by the Sun, which exacts pronounced responses to any variations in solar spectral irradiance and incoming atomic particles. Examples can be found in the dramatic solar-driven changes in the number of electrons and ions in the ionosphere, and in the instantaneous temperature responses of the thermosphere. Variations that are clearly related to the 11-year solar cycle are also apparent in the lower stratosphere, in the long series of readings taken there by sounding balloons.

All of these more rarefied regions of the Earth's atmosphere are more directly exposed and in many ways more vulnerable to solar disturbances and fluctuations. But the dense troposphere—the realm of all weather and climate—is only weakly linked to the more vacuous middle and upper atmosphere, due to these great differences in density.

Apparent 11-year effects have been noted through the years in local and regional weather records, but it is only recently that the marks of the Sun have been unequivocally found in hemispheric or globally-averaged tropospheric data sets.

One of these, covering forty years of Northern Hemisphere temperature measurements from balloon sondes at heights of about two to eight miles above the surface, reveals a clear 11-year signal, in phase with the solar activity cycle, with higher temperatures, as expected, in years of maximum solar activity. Another, covering the entire globe, found a similar 11-year temperature response that varied, in both sign and amplitude, as a function of both latitude and height above the surface. The nature of the differences suggests that the solar signal observed is imposed from the stratosphere, as a result of dynamical motions in the atmosphere.

Another—cited earlier—based on global measurements of the troposphere in low and mid latitudes from 1958 through 2001 found that not only the temperature, but all major meteorological observables in low and mid-latitudes were strongly correlated with the phase of the 11-year solar cycle, when signals due to other known sources of climate forcing were removed from the data. The latter included known El Niño Southern Oscillation (or ENSO) effects, volcanic eruptions, changes in atmospheric aerosols, and a linear trend attributed to global greenhouse gases.

Solar Forcing of the Oceans

The oceans are a major element of the global climate machine, affecting year-to-year and longer changes in climate through ocean-atmosphere interactions and internal modes of climate variability such as the Pacific Ocean ENSO and the North Atlantic Oscillation, or NAO: a similar phenomenon involving a quasi-regular see-saw in sea-level air pressure between large-scale regions, in this case in the North Atlantic Ocean.

The oceans also influence climate by storing heat in ocean basins, and by temperature-driven changes in the thermohaline circulation of the deep oceans that are driven by systematic differences in the salinity of ocean water.

Variations in solar irradiance are a likely forcing factor in each of these, as has been shown in recent studies of different ocean characteristics. In one of them, 11-year cyclic variations were found in surface and subsurface ocean temperatures, which were consistent—in both phase and amplitude—with measured 11-year variations in solar irradiance. Another, which we describe later, found a striking correlation throughout the last 11,000 years between ocean temperatures and recurrent eras of prolonged suppression of solar activity, like the Maunder Minimum.

Clear marks of 11-year solar forcing were also found in ocean temperature measurements covering forty-two years of surface and sub-surface temperature sampling in the Atlantic, Indian, Pacific, and global-averaged Oceans, spanning conditions from latitude 20° to 60° N. These reveal systematic cyclic changes in ocean temperatures, like those found in the atmosphere, that are in phase with the 11-year solar activity cycle, and consistent with the variations found in total solar irradiance. The 11-year temperature response was evident from the ocean surface to a depth of about 500 feet, which is about the maximum distance to which sunlight penetrates. (The transparency of the photic zone of the upper ocean—which in the clearest ocean water extends to a depth of about 650 feet—is greatest in the blue part of the visible spectrum, and in more turbid seashore waters in the green and green-yellow. To most of the ultraviolet and infrared, the seas are almost opaque.)

Although many questions remain, these documented changes in long-term records suggest that natural modes of the global climate system are locked in phase to the 11-year solar activity cycle.

That subtle changes in solar irradiation might serve as a phase-locking device for year-to-year changes in climate is in agreement with our present understanding of the glacial-interglacial cycles, tens of thousands of years long, that characterize the climate of the Pleistocene epoch of the last million years or so. In the latter case, relatively small changes in the distribution of insolation (radiant solar energy) over the globe, arising from gradual changes in the orbit of the Earth and the inclination of its axis of rotation—today a tilt of 23½°—are thought to have served as the pace-maker for the coming and going of the major Ice Ages.

Hidden Diaries of the Ancient Sun

Radiocarbon (^{14}C) is an unstable isotope of carbon that has long been employed in archaeology for establishing the date of production of samples of cellulose or bone or other carbon-bearing matter. It and the beryllium isotope of atomic weight ten (^{10}Be) are the best known and most abundant of the so-called cosmogenic nuclides which when naturally sequestered in wood or compacted snow can also serve as indirect, or proxy indicators of past changes in solar activity. The recovery and analysis of both of these—^{14}C from the wood in annual growth-rings of the oldest-living trees, and ^{10}Be in layered ice drawn from deep polar cores—have confirmed the existence of prolonged episodes of major depression in the overall level of solar activity, such as the Maunder Minimum of 1645-1715. In so doing they have greatly extended the span of retrievable solar history.

Cosmogenic nuclides are created in the rarefied air of the middle atmosphere when energetic cosmic rays impinge on neutral atoms of air. Each carbon-14 atom is the product of a direct hit by an incoming neutron—itself the newly-born product of a preceding cosmic ray collision higher in the atmosphere—on an atom of neutral nitrogen (^{14}N): the most abundant component of air.

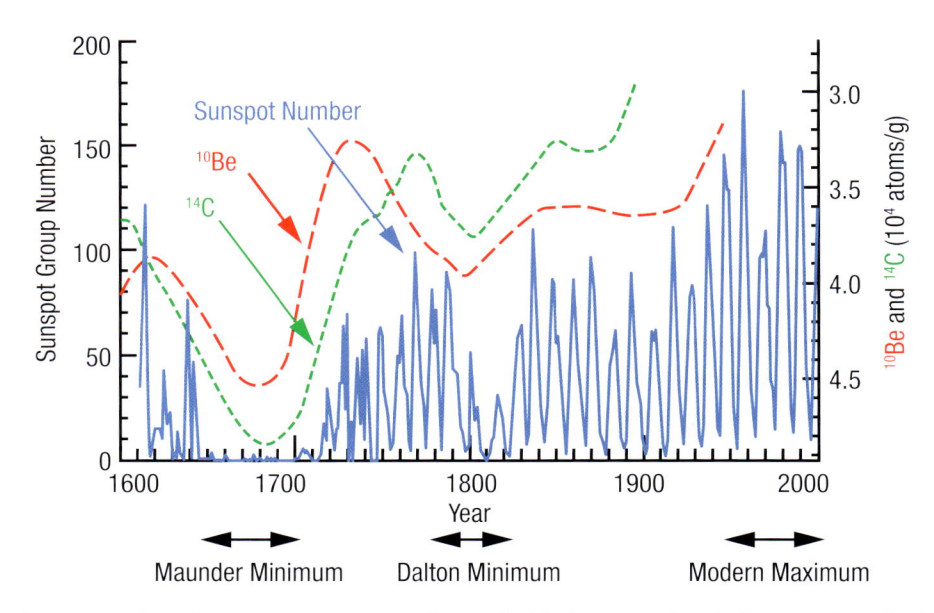

Average number of sunspot groups seen each year (in blue) compared with the rates of production of radiocarbon(^{14}C, dashed green) and beryllium-10 (^{10}Be, dashed red) in the Earth's atmosphere for the period since the introduction of the telescope on 1610. These cosmogenic isotopes are produced by incoming cosmic rays whose numbers are modulated by solar activity, and hence serve as indirect proxies of solar activity. Their past concentrations are obtained for carbon through laboratory analyses of dated tree-rings and for beryllium from cores of layered Arctic and Antarctic ice. The data used for carbon-14 measurements come to an end at about 1900, by which time the carbon introduced into the atmosphere by the ever-increasing burning of fossil fuels erases, in effect, the more subtle solar signature in the tree-ring record.

In the process a proton is jarred loose from the impacted atom's nucleus, leaving behind an unstable isotope of carbon of atomic weight 14: the most abundant cosmogenic nuclide. ^{10}Be, the second most abundant, is produced by a similar process when incoming cosmic ray protons bombard and break apart neutral atoms of either oxygen or nitrogen.

The number of cosmic ray particles that reach our planet to do this work is modulated by changes in both the strength of the Earth's magnetic field, and by changes in solar activity, both of which act to deflect cosmic rays that pass through the heliosphere.

The strongest of these are changes on times scales of thousands of years in the strength of the Earth's magnetic field. The higher frequency (10 to 100 yr) effects of solar changes are imposed on top of these like noise on a sinusoidal wave form. Among these higher frequency effects are the slower but greater amplitude changes like the Maunder Minimum in the overall level of solar activity, persisting for decades to a hundred years or so. Present as well are the even higher frequency effects of the Sun's eleven-year cycle of activity.

More cosmic particles reach us when the Earth's magnetic field is weaker, as it was seven thousand years ago, and fewer as it grows stronger, as it did until about two thousand years ago when it began again to gradually weaken.

More cosmic particles also reach the Earth when the Sun is less active; and fewer when it is more active. Thus the rates of production of ^{14}C and other cosmic nuclides in our atmosphere are strongly *anti*-correlated with solar activity, as confirmed by more than half a century of direct observation of the flux of cosmic rays using neutron monitors at mountain-top stations.

The Fate of Carbon-14

Soon after they are created, atoms of ^{14}C combine with neighboring atoms of oxygen to form gaseous carbon dioxide, most of which, in time, diffuses downward into the lower atmosphere. When it reaches ground level, some is ingested through pores in the leaves of trees.

There, through photosynthesis, CO_2 separates into its component atoms of carbon and oxygen. The freed carbon, still of weight 14, is then metabolized in the plant, this time combining with hydrogen and oxygen to form cellulose: the carbohydrate that is a fundamental constituent of all plant fiber, including the wood of trees.

And so it is that a carbon atom of weight 14, born of mixed parents high in the sky, is—after years of traveling around the world—at last sequestered in the thin sheath of living cells that form a ring of springtime growth in the stem of a growing tree.

☉   ☉   ☉

Carbon 14 is inherently unstable in that it radio-actively decays with time into ordinary carbon of weight 12.

After 5730 years—the expected half-life of the isotope—about 50% of the ^{14}C remains, and after 30,000 years, only 10%. When a third of a million years has gone by, 99% of the original ^{14}C has been converted to ordinary ^{12}C. In coal and oil and natural gas—which were created during the late carboniferous period, about 300 million years ago—all of the radiocarbon then present in the small aquatic plants and animals that gave us these now troublesome fuels is gone.

Thus, like sand in an hourglass, the amount of radiocarbon present in any form of carbon-based life—living or dead—tells us of its age: or more specifically, the time that has gone by since the instant each atom of radiocarbon was created, high in the sky. To a close approximation this tells as well of the time that has elapsed since the plant-eating animal whose bones we find was last alive and grazing; or when the tree from which the sample of wood was cut, however long ago, was yet in leaf.

⊙   ⊙   ⊙

The ratio of ^{14}C to ^{12}C in a dated tree-ring also reflects the average *rate of production* of ^{14}C in the atmosphere, which is in turn a measure of the flux of high-energy cosmic rays at the Earth, and from this, the level of solar activity at that time.

Through this chain, precision analyses of ^{14}C/^{12}C ratios in annual rings of the long-lived bristlecone pine have provided a continuous, proxy record of long-term changes in the level of solar activity through the last 11,000 years—since the end of the last Ice Age—extending the length of the historical, written record of sunspots by a factor of about thirty. An 11-year modulation, though severely attenuated by the amount of time it takes for the newly-created carbon-14 atom to make it into the leaves of trees, can be detected, albeit with difficulty, throughout. A much more obvious feature in the long radiocarbon record are repeated Maunder Minimum-like depressions in the overall level of solar activity, each persisting for thirty to about 100 years.

Beryllium-10 in Ice Cores

An independent verification of these reconstructions of solar history is available through the analysis of another cosmogenic nuclide, beryllium-10 (^{10}Be), sequestered in polar ice and deep-ocean cores. Since the rates of deposition of ^{10}Be and of ^{14}C result from very different processes, we can at first assume that major features common to both of them are due to their rate of creation, and hence traceable to the Sun.

Cosmogenic ^{10}Be is like ^{14}C in that it accumulates in the lower stratosphere for a while before entering the troposphere. But unlike radiocarbon, beryllium-10 finds its way down to the Earth's surface more directly and expeditiously, through either precipitation or dry deposition, and is not biogeochemically recycled en route as carbon-14 often is. Although we never see or feel it, ^{10}Be is deposited on our roofs and yards and the tops of our cars, wherever we are, each time it rains or snows.

⊙   ⊙   ⊙

in ocean temperatures, like those found in the atmosphere, that are in phase with the 11-year solar activity cycle, and consistent with the variations found in total solar irradiance. The 11-year temperature response was evident from the ocean surface to a depth of about 500 feet, which is about the maximum distance to which sunlight penetrates. (The transparency of the photic zone of the upper ocean—which in the clearest ocean water extends to a depth of about 650 feet—is greatest in the blue part of the visible spectrum, and in more turbid seashore waters in the green and green-yellow. To most of the ultraviolet and infrared, the seas are almost opaque.)

Although many questions remain, these documented changes in long-term records suggest that natural modes of the global climate system are locked in phase to the 11-year solar activity cycle.

That subtle changes in solar irradiation might serve as a phase-locking device for year-to-year changes in climate is in agreement with our present understanding of the glacial-interglacial cycles, tens of thousands of years long, that characterize the climate of the Pleistocene epoch of the last million years or so. In the latter case, relatively small changes in the distribution of insolation (radiant solar energy) over the globe, arising from gradual changes in the orbit of the Earth and the inclination of its axis of rotation—today a tilt of 23½°—are thought to have served as the pace-maker for the coming and going of the major Ice Ages.

Hidden Diaries of the Ancient Sun

Radiocarbon (^{14}C) is an unstable isotope of carbon that has long been employed in archaeology for establishing the date of production of samples of cellulose or bone or other carbon-bearing matter. It and the beryllium isotope of atomic weight ten (^{10}Be) are the best known and most abundant of the so-called cosmogenic nuclides which when naturally sequestered in wood or compacted snow can also serve as indirect, or proxy indicators of past changes in solar activity. The recovery and analysis of both of these—^{14}C from the wood in annual growth-rings of the oldest-living trees, and ^{10}Be in layered ice drawn from deep polar cores—have confirmed the existence of prolonged episodes of major depression in the overall level of solar activity, such as the Maunder Minimum of 1645-1715. In so doing they have greatly extended the span of retrievable solar history.

Cosmogenic nuclides are created in the rarefied air of the middle atmosphere when energetic cosmic rays impinge on neutral atoms of air. Each carbon-14 atom is the product of a direct hit by an incoming neutron—itself the newly-born product of a preceding cosmic ray collision higher in the atmosphere—on an atom of neutral nitrogen (^{14}N): the most abundant component of air.

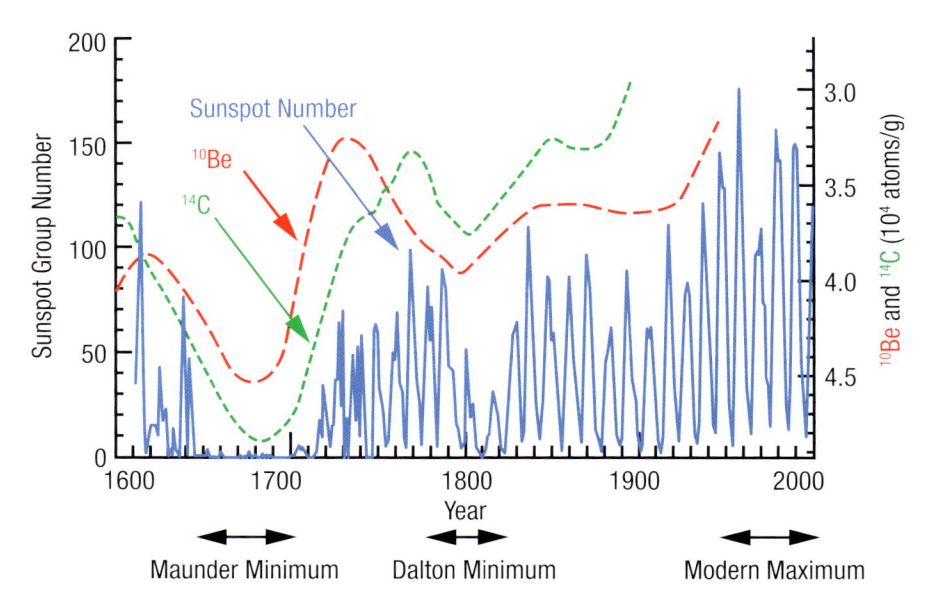

Average number of sunspot groups seen each year (in blue) compared with the rates of production of radiocarbon (^{14}C, dashed green) and beryllium-10 (^{10}Be, dashed red) in the Earth's atmosphere for the period since the introduction of the telescope on 1610. These cosmogenic isotopes are produced by incoming cosmic rays whose numbers are modulated by solar activity, and hence serve as indirect proxies of solar activity. Their past concentrations are obtained for carbon through laboratory analyses of dated tree-rings and for beryllium from cores of layered Arctic and Antarctic ice. The data used for carbon-14 measurements come to an end at about 1900, by which time the carbon introduced into the atmosphere by the ever-increasing burning of fossil fuels erases, in effect, the more subtle solar signature in the tree-ring record.

In the process a proton is jarred loose from the impacted atom's nucleus, leaving behind an unstable isotope of carbon of atomic weight 14: the most abundant cosmogenic nuclide. ^{10}Be, the second most abundant, is produced by a similar process when incoming cosmic ray protons bombard and break apart neutral atoms of either oxygen or nitrogen.

The number of cosmic ray particles that reach our planet to do this work is modulated by changes in both the strength of the Earth's magnetic field, and by changes in solar activity, both of which act to deflect cosmic rays that pass through the heliosphere.

The strongest of these are changes on times scales of thousands of years in the strength of the Earth's magnetic field. The higher frequency (10 to 100 yr) effects of solar changes are imposed on top of these like noise on a sinusoidal wave form. Among these higher frequency effects are the slower but greater amplitude changes like the Maunder Minimum in the overall level of solar activity, persisting for decades to a hundred years or so. Present as well are the even higher frequency effects of the Sun's eleven-year cycle of activity.

Thus, had he but known, Robert Louis Stevenson in *A Child's Garden of Verses* could instead have written

> *The rain is raining all around,*
> *It falls on field and tree.*
> *Big drops of H_2SO_4*
> *And chunks of ^{10}Be.*

The beryllium that reaches the surface in polar regions is entrapped in annual layers of fallen snow that are in time compressed and preserved as annual layers of hardened ice, initially separated by lines of superficial thawing and refreezing. Since ^{10}Be is rapidly washed out of the air by precipitation, its abundance in the atmosphere varies considerably in both space and time. Because of this, the amount sequestered in any ice core layer reflects not only the solar-driven production rate of the cosmogenic nuclide but also atmospheric circulation patterns and precipitation rates that brought it to that individual spot on the ground in the snows of yesteryear.

Because of this and unlike ^{14}C—which in the form of CO_2 is uniformly mixed throughout the global atmosphere—the amount of ^{10}Be that accumulates on the surface of the Earth varies considerably from place to place: very much a function of the amount of rain or snow that has fallen there, and hence of local and regional weather. Variations found in layers of snow and ice at any site can be corrected for local precipitation effects, but corrections for non-local meteorological variations in the delivery of ^{10}Be to the ice caps is far more difficult.

Nevertheless, sequestered in a much older repository, and with a half-life of 1.5 million years (compared to 5730 years for ^{14}C) ^{10}Be in ice offers the potential of extending the reach of recorded solar history in the deepest, Greenland ice cores more than 200,000 years into the past, and potentially, in the deepest Antarctic cores, to as much as 400,000 years, through a number of glacial-interglacial cycles: in all, a span of time that tody is the best documented and most studied period of the past climate of the Earth.

Since the ^{10}Be record is capable of finer temporal resolution than ^{14}C, what may be more valuable is the prospect of high resolution year-by-year information on changes in solar activity through the last several thousand years, derived from analyses of ice nearer the surface, where annual layering is more clearly preserved.

As one drills deeper and deeper into the ice, what were once clearly-defined annual layers of compressed snow and firn—separated by annual lines of partial thawing and refreezing—gradually lose these valued earmarks: their identities squeezed out of them by the accumulating load of glacial snow and ice that lay above, in some places more than a mile deep.

Still, what these ice-core or tree-ring proxies tell us of the ancient Sun is not a direct account of past changes in solar *irradiation*—which climatologists would most like to know—but of changes in solar *magnetic activity*, which is one step removed.

Marks of the Sun on North Atlantic Climate During the Last 11,000 Years

An extensive paleo-oceanographic study—based on the recovery of data that tell of climatic changes during the present post-glacial or [Holocene epoch](#)—has yielded what may be the most compelling evidence for a connection between longer-term changes in climate and the changing moods of the Sun. Found was not one but an unbroken series of responses of regional climate to episodes of suppressed solar activity like the Maunder Minimum, each lasting from 50 to 150 years.

The paleoclimatic data, which cover the full 11,000 year span of the present interglacial epoch, were derived from records of the concentration of identifiable mineral tracers in layered sediments on the sea floor of the northern North Atlantic Ocean. The timing and duration of periods of suppressed solar activity came from a different source: from analyses of the ratio of ^{14}C to ^{12}C in dated growth rings of the long-lived bristlecone pine that cover the same period of time, and from concurrent measurements of ^{10}Be in polar ice cores.

The minerals used as ocean tracers come from the soil in high latitude regions, which, through erosion, is carried out to sea in drift ice. When southerly-drifting ice reaches waters warm enough to melt it, the tracers are released and sink to the bottom of the ocean. There they are gradually covered over with other sedimentary material, that in time build up a layered history of what fell to the bottom of the ocean at that place, and when.

Sea-cores drilled from ships at sea into the mud of the ocean bottom can sample this imbedded information. Each core is in effect a time capsule that holds a continuous record of the material that has sunk to the sea-floor at that location. Analyzing many cores in this way—taken from different locations in the North Atlantic Ocean—allows the investigator to draw a meaningful map of the southern limit of drifting sea ice at any time in the past.

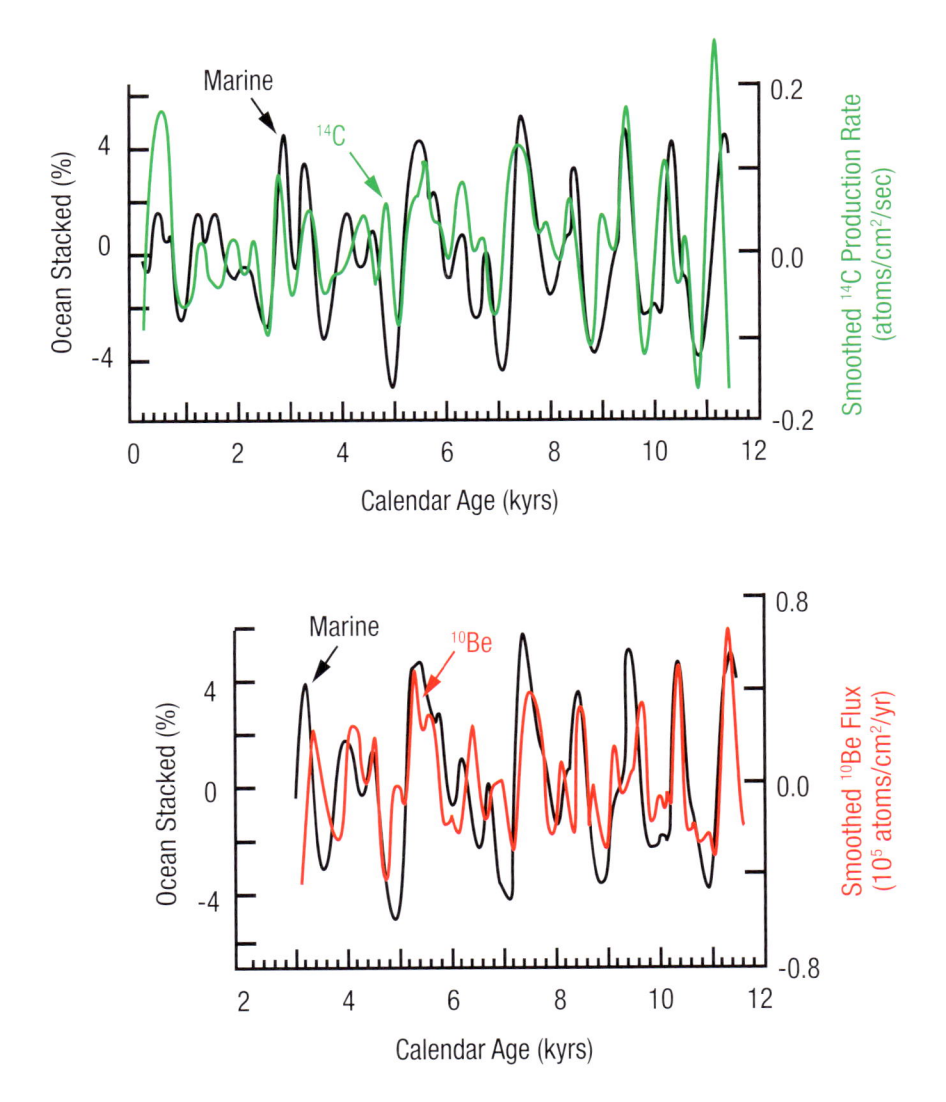

Comparison of proxy records of solar activity from tree-rings (upper figure, in green, for carbon-14) and ice cores (lower, in red, for beryllium-10) with temperature in the North Atlantic ocean (black curve, upper and lower, from deep-sea cores) for about the last 11,000 yrs (the Holocene epoch). Evident is a remarkable agreement between long-term excursions in climate, lasting hundreds of years, and contemporaneous conditions on the Sun. Periods of suppressed solar activity (peaks in the red and green curves) like the Maunder Minimum correspond to times of cooler ocean temperatures, with higher temperatures during long periods of greater solar activity which is of the same nature as the suggested connection in more modern times between the Maunder Minimum in solar activity (1645-1715) and the coldest temperatures of the concurrent but longer-lasting Little Ice Age.

In years when the climate is systematically colder in the region, waters sufficiently warm to melt floating ice are not encountered until the floe has drifted far to the south. In warmer periods, the opposite is true, and the southern limit of drift ice shifts back toward polar waters.

The study revealed that the sub-polar North Atlantic Ocean has experienced nine distinctive expansions of cooler water in the past 11,000 years, occurring aperiodically but roughly every 1000 to 2000 years, with a mean spacing of about 1350 years.

Each of these regional cooling events coincides in time with strong, distinctive minima in solar activity, derived from records of the production of ^{14}C from tree-ring records and of ^{10}Be from deep-sea cores covering the same period of time.

This remarkable North Atlantic finding—the patient and painstaking work of a diverse group led by the late Gerard Bond—suggests that solar variability has been a major climate driver throughout the present Holocene epoch.

But to produce so strong and consistent a response, the solar driving force—whatever its origin—would need some assistance, if what we know of the bounds of solar variations still applies to these longer time scales. One of these might be an amplification of the impacts of slowly-varying changes in solar irradiation through their impact on the oceans' thermohaline circulation.

An envisioned astronaut on the surface of Mars, with actual images of the Moon, Earth, and Sun placed in the Martian sky, shown here much larger than they would actually appear from that distance.

FORECASTING SPACE WEATHER AT THE EARTH AND BEYOND

Space Weather

The terms "space weather" and "space climate" were coined not long ago to describe current and time-averaged conditions in the Earth's outer *environment*, in the same way that "weather" and "climate" refer to current or time-averaged conditions in the lower atmosphere. Space weather includes any and all conditions and events on the Sun, in the solar wind, in near-Earth space and in our upper atmosphere that can affect space-borne and ground-based technological systems and through these, human life and endeavor.

As noted earlier, the economic impacts of severe space weather events—such as the major electric power outages that can follow strong geomagnetic storms, or the disabling effects of solar eruptions on costly spacecraft—can easily reach the level of direct and indirect losses now associated with the most severe weather phenomena. Even moderate space weather events can affect national security, and in many ways. Extreme space weather events can very seriously also affect human life and health on extended lunar or planetary missions of exploration.

☉ ☉ ☉

Terrestrial weather has long been of interest and concern to everyone on Earth. But in the modern, interconnected high-tech world of today, *space* weather is fast becoming equally important and down-to-earth. Today it can affect anyone who has a television set, a radio, or a computer, or who makes use of GPS in any way; all who travel on jet aircraft flights that cross high-latitude regions in either hemisphere; nearly every city, large or small, that is linked to an electric power grid; every ship afloat; every spacecraft that is sent into the sky; and any nation whose security rests in part or all on radio communications, radar, or space-borne equipment.

In predicting day-to-day *weather* the volume of space in which measurements are taken and predictive models are run is limited to the troposphere: a thin veneer of air extending no more than about seven miles above the surface of the Earth.

In contrast, the vast domain of space weather and space climate starts in the upper atmosphere—about fifty miles above our heads, and reaches all the way to the Sun itself: nearly 100 million miles away. Within this vast region of interest are the ionosphere, thermosphere, magnetosphere and radiation belts; the immense volume of interplanetary space through which high-energy solar and cosmic particles pass to reach the Earth; the Moon and the inner planets, and the fiery furnace of the Sun itself.

Predictions

We have little if any control over the weather, and none at all over space weather events. But we can minimize the societal impacts of variations in either of them through reliable predictions of what is expected—or not expected—at any time.

Forecasts of the next day's weather at any place customarily include high and low air temperatures, precipitation, cloud cover, expected winds, and other meteorological parameters derived from computerized models of atmospheric circulation at different heights in the troposphere. The data needed to supply these models come largely from *in situ* measurements and observations obtained from a shared, world-wide network of weather stations; from balloon sondes; and since 1975—when the first *Observational Environmental Satellite (GOES-1)* was put into orbit—from direct, large-scale pictures of cloud cover and storm tracks seen from a height of 22,200 miles.

Information regarding space weather is obtained in much the same way, as are corresponding three-day or longer forecasts of things to come.

Among the need-to-know users of reliable *space weather* information are commercial airlines; electric power companies; almost every segment of the telecommunications industry including cell-phone and GPS providers; law enforcement; homeland security; banking and other commercial interests; those who go down to the sea in ships; all branches of the military; NASA and other national or international space agencies, particularly at times of human or robotic exploration; and commercial space and space tourism ventures.

Meeting the specific needs of these and other users often calls for tailored "predictions" of various kinds, all of them based in large part on computerized models that reproduce the behavior of parts or all of the Sun-Earth system.

The most common prediction is a "forecast": a prediction issued before an event such as flare or geomagnetic storm takes place. An example is the arrival of a fast moving CME, or a severe magnetic storm. But there are useful predictions of other kinds.

A "nowcast," as in meteorology, is a statement of current conditions at this time. Actual "predictions" may be in the form of an "alert," which might state, for example, that *A large fast-moving CME appears at this time to be headed toward the Earth;* a "watch" telling for example whether it is likely (or unlikely) that the present very low or very high level of solar activity will persist through the next two weeks; or a "warning" such as *Major disruptions in the ionosphere are expected during the next eight to ten days, in connection with expected CME and flare activity.*

⊙ ⊙ ⊙

In the past, warnings of unfavorable weather and impending storms on land or sea were often issued using color-coded lights or flags, not unlike the red-orange-yellow advisories that were a few years ago tried to signal different levels of perceived homeland security risk.

In Birmingham, Alabama for example, a continuously-burning electric beacon was for many years employed to alert residents of potentially hazardous driving conditions. The warning light—installed where all could see it, atop an iron statue of Vulcan, 55 ft. tall and set on a high pedestal at the summit of Red Mountain—burned *green* for "all clear" or *red* when a fatal traffic accident had occurred.

Simple warning systems of a similar nature employing displays of colored flags and lanterns were and are still used in marinas, harbors and many coastal points to advise mariners or recreational sailors of expected hurricanes and other severe storm conditions.

Today and around the world, warnings of severe weather—at sea, on land, or in space—are broadcast or disseminated electronically and instantly to anyone concerned; and in more detail, which is required to serve the specific and often tailored needs of the many different users of this information. This is particularly true in space weather, where needless to say, a simple color-coded alert system—perhaps *red* to indicate that solar or geomagnetic storms are expected, *green* that they are not, and *orange* and *yellow* for different degrees of ambiguity—would not be of much use in this case, either.

USEFUL SPACE WEATHER PREDICTION
THE SUN   • Probability, magnitude and expected duration of solar flares   • Probability, size, speed and direction of CMEs   • Probability and nature of solar energetic particle (SEP) events   • 27-day or longer forecasts of solar conditions
INTERPLANETARY AND NEAR-EARTH ENVIRONMENT   • Intensity of solar EUV irradiation   • Particle radiation environment in near-Earth space   • Solar wind plasma parameters: density, velocity and magnetic field orientation   • Trajectories, predicted arrival time and likely impact of CMEs   • Cosmic ray flux level
MAGNETOSPHERE AND RADIATION BELTS   • Expected conditions in the magnetosphere and radiation belts   • Strengths of electric current systems in the magnetosphere   • Expected geomagnetic storms including onset, intensity, and duration   • Ground induced currents
UPPER AND LOWER ATMOSPHERE   • Conditions in the thermosphere and ionosphere   • Onset, intensity and expected recovery of ionospheric storms   • High energy particle fluxes at jet aircraft altitudes

Sources of Needed Data

In some instances the information needed for an accurate space weather forecast are secured—as in meteorology—by direct, *in situ* sampling. Examples are data regarding the composition, energy and magnetic field orientation of the solar wind plasma, sampled when possible by spacecraft that are at the time immersed in that medium. Another is the continuous record of characteristics of the Earth's magnetic field recorded by magnetometers through a global network of ground-level stations. A third is *in situ* measurements of the ambient flux of high-energy particles sensed by the radiation counters routinely carried for determining the dosage of ionizing radiation received on some European airlines.

But most of the data that provide current information about the Sun, cosmic rays, the magnetosphere and plasmasphere, and the thermosphere and ionosphere are arrived at indirectly, or observed from afar by remote sensing.

Information regarding the Sun comes chiefly from telescopic observations made either on the ground or by spacecraft above the atmosphere. Images of the corona, needed for the prediction and monitoring of CMEs come mostly from space-borne x-ray telescopes and visible-light coronagraphs. The total and spectral irradiance received from the Sun—from the x-ray and EUV to the infrared—is also continuously monitored from afar by space-borne radiometers designed to secure these critical data from above the interfering atmosphere.

The flux of solar EUV radiation obtained from solar-pointed spacecraft is a particularly vital source of information because of its role in configuring the layered ionosphere, and heating the thermosphere. Estimates of the flux of solar EUV irradiation—and its variation from minute to minute or year to year—are also obtained, though less directly, from ground-based solar radio telescopes that monitor the intensity and variability of coronal radio emission. Ground-based measurements of the radiation of the unresolved disk of the Sun in two narrow spectral lines—that of ionized calcium in the near-ultraviolet and helium in the near-infrared—provide other proxies for the whole-disk solar EUV radiation.

When solar radio antennas are pointed at the Sun and their receivers tuned to 2.80 gigahertz (10.7 cm wavelength) in the ultra-high frequency band, they receive solar radio emission that comes from the same level in the Sun's outer atmosphere that produces energetic solar EUV radiation. Because of this, records of solar radio emission at this wavelength and of whole-disk EUV radiation closely track each other.

Like measurements from space of the total solar irradiance in the EUV, 10.7 cm radio emission from the Sun records the *averaged* radiation from the entire visible disk, without spatially resolving individual bright or dark features: as though the Sun were a star-like dot in the sky. Because of this built-in averaging, measurements of 10.7 cm solar radio emission have since the 1950s provided an alternate spatially-averaged index of solar activity that complements the daily sunspot number.

Of these two, the 10.7 cm radio index is less subjective than the venerable sunspot number, which was arbitrarily defined by Rudolph Wolf in Zürich in 1849 as the number of discernible spots seen on the disk plus ten times the number of sunspot groups, all multiplied by an estimated correction factor to compensate for perceived differences in observers, telescopes, observing sites and atmospheric conditions.

☉ ☉ ☉

An indirect measure of the flux of galactic cosmic rays at the top of the Earth's atmosphere is provided by neutron monitors operated at high-altitude stations around the world. The numbers of neutrons that penetrate to these lower levels is at any time a proxy indicator of the incidence of primary cosmic rays higher in the atmosphere, which through collisions with atoms and molecules of air produce these most energetic secondary cosmic ray particles.

The density of electrons in different layers of the ionosphere—the determining factor in forecasts of radio propagation conditions—is also recovered indirectly by remote sensing, in this case employing radio waves transmitted upward from the ground. As in conventional radar, the strength and timing of the reflected signal can be read in terms of the local structure and electron density in the layered atmosphere, fifty to several hundred miles overhead.

Available Warning Times

The extreme speeds at which energetic solar particles and radiation stream outward from the Sun impose severe limits on the amount of time available, once an event is observed there, to react to or mitigate its possible effects.

All electromagnetic radiation leaves the Sun at the same velocity—the speed of light, about 186,000 miles per second—and reaches the Earth in eight minutes. The speeds of atomic *particles* depend upon their own energy, at the Sun and en route.

Slow-speed streams in the solar wind can take more than four days to make the journey, while solar energetic protons (SEPs) can arrive in a matter of tens of minutes or less. Less energetic particles from the Sun, and those carried outward in CMEs, generally arrive at the orbit of the Earth in from one to three days.

But the warning time available to us on Earth is always eight minutes less than the time it takes particles or photons to get here.

This automatic deduction is a consequence of the fact that what we see when we look at the Sun is not what is happening at this minute, but what happened there eight minutes ago: the time it takes visible light to travel the 93 million miles that separate us from the star. Were we on Mars, a look at the Sun would show us what it was like there about 12 minutes before; and on Jupiter, almost three quarters of an hour ago.

TRAVEL TIME FROM THE SUN TO THE EARTH

PARTICLES	SPEED IN MILES/SEC	TIME TO REACH THE EARTH
Particles in the slow solar wind	250	4½ days
Particles in a typical CME	250	4½ days
Particles in high-speed streams	470	2½ days
Less energetic particles from solar flares	360 to 1100	1 to 3 days
Particles in the fastest CMEs	1200	30 minutes to several hours
Energetic protons from solar flares	25,000 to 100,000	15-60 minutes
ELECTROMAGNETIC RADIATION		
X-ray, EUV, UV, Visible, Infrared, and Radio radiation	186,000	8 minutes

Thus the moment we first see the bright flash of a solar flare, the energetic EUV and x-ray radiation it sends our way has already arrived—as though it got here instantly; and the damaging high energy protons that can travel to the Earth in but 15 minutes time are already more than halfway here, leaving us at most but seven minutes' warning time. For most other particles—including the fastest plasma in CMEs, the obligatory eight-minute deduction is not as significant.

☉ ☉ ☉

How much warning time is needed to mitigate possible deleterious effects at the Earth?

An hour of advanced warning of a severe geomagnetic storm can be helpful for electric power companies, for in that time they can change the way they generate and distribute energy. An hour or less may be enough to allow air controllers to reduce the altitude of flights passing through polar regions or to redirect the flight paths of those en route. A half hour to an hour *may* be enough for astronauts engaged in extravehicular activity to scurry back inside, or for those on the surface of the Moon or Mars to seek shelter—if there is any—from an anticipated burst of extremely energetic particles from the Sun.

☉ ☉ ☉

There are also ways of gaining more warning time.

One is to delay the mission itself if unusual solar activity is expected, or to postpone scheduled space walks en route to or on the surface of the Moon or Mars. Another is to direct crews to hunker down in parts of the command module, or if space allows, in the lunar lander that may in places offer marginal protection.

The most beneficial and challenging way to increase the warning time is through reliable predictions of when and where likely solar eruptive events will occur. For CMEs, the prediction must reliably foretell whether the expanding blob of plasma will cross the Earth's path. If so, its size and speed and the orientation of the magnetic field it carries are also needed, as is its projected time of arrival and the probability of intense geomagnetic storms.

Alerts and warnings must first of all be reliable, since in many applications false alarms can lead to considerable unnecessary expense. Examples are electric power companies who will reconfigure power distribution networks or put on extra equipment in anticipation of a major geomagnetic storm; airlines who will expend additional fuel and delay or cancel flights; telecommunications carriers who will shift frequencies and strategies. False alarms can also limit the accomplishments and ultimate success of costly space missions when tightly-scheduled activities are cancelled, or result in a depletion of orbital control fuel.

Assistance in increasing the accuracy and the warning time of forecasts is now available thanks to the deployment of early-warning spacecraft that are stationed, like sentinels, nearer to the Sun than we. In these advanced positions they can send back vital information, as from scouts sent ahead of an advancing troop of the U.S. Cavalry.

Radioed early warnings sent back to us from distant sentinels can never outrun the x-ray and EUV photons that race past them, traveling at the speed of light toward the Earth. But they will get here well in advance of approaching CMEs, fast streams in the solar wind, and most energetic particles coming from flares.

The *ACE* spacecraft, launched in 1997 and still hard at work, was designed to study the composition of plasma and energetic particles in the heliosphere from a distant outpost nearly a million miles closer to the Sun than we, where it serves as a watchman or sentinel of this kind.

Especial Needs for Manned Space Exploration

Streams of solar protons and heavier ions with particle energies in the range of 10^6 to 10^9 electron volts can pose severe hazards for manned space missions that venture beyond the natural protection of the Earth's atmosphere and magnetosphere. Continuous exposure to galactic cosmic rays, with even higher energies is equally if not more hazardous to manned space flight. Exposure to sufficient doses of the ionizing radiation of highly-energetic atomic particles can damage cells in living tissue and organs that provoke cellular mutations and incipient cancer; induce nausea and the debilitating symptoms of radiation sickness that were felt by victims of atomic bomb explosions sixty years ago; and lead to slow or sudden death.

At the same time, each high-energy atomic particle that streams through our bodies is not a deadly bullet. Nor is it likely that a single burst or stream of them would be.

The critical factor—for astronauts, pilots, passengers, or patients in a dentist's chair—is, as noted earlier, the *accumulated dosage* that one receives: the product of the *intensity* of the ionizing radiation (determined by the speed, mass, and number of ionizing particles) and the amount of time one is exposed to it.

☉   ☉   ☉

Every manned spacecraft that ventures beyond the Earth's atmosphere and magnetosphere will find itself caught in the crossfire of energetic particles coming from as many as four different sources.

The first and least hazardous is the ubiquitous solar wind in which spacecraft are immersed as soon as they leave the protective shields that surround the Earth: the ever present flow of low energy (1 – 10 keV) particles borne outward from the Sun in a turbulent mix of steady breezes and sudden gusts. Immensely more energetic cosmic rays are also ever present, arriving from every direction in the sky and able to pierce almost anything.

A spaceship in deep space will at times also find itself directly in the path of fast-moving CMEs: but these particles, too, are limited to the low energy range of a few KeV and are not ordinarily a direct threat to space travel.

The least frequent but probably most hazardous to life and health are the solar energetic protons (SEPs) and heavier ions that are accelerated outward by CMEs and major solar flares, and in the shock waves that form—like the bow wave of a ship—some distance ahead of the fast-moving plasma. These travel toward us at almost unimaginable velocities which can approach the speed of light. When coupled with their appreciable mass, this makes SEPs, with cosmic rays, the principal threats to be avoided in manned space flight. And they can rain down on our planet, on the Moon and Mars and in near-Earth space for hours on end.

The challenge in deep-space exploration of the Moon and Mars is to provide shielding sufficient to block or deter any of these ultra-high-energy particles from reaching parts of the spacecraft where astronauts live, work or seek shelter.

It has been estimated that astronauts on extended trips like those envisioned for Mars or for manned colonies on the Moon would each year receive from galactic cosmic rays fully 16 times the maximum authorized dose of ionizing radiation prescribed for nuclear plant workers. Moreover, to be effectively shielded from GCRs while on the surface of either body would require burying a lunar or Martian base below hundreds of tons of lunar or Martian soil.

Sunset at the close of a Martian day seen from the Mars Pathfinder on the planet's surface. With a much thinner atmosphere—evident here in the absence of sunset colors—almost no UV-absorbing oxygen, and no magnetic field, visitors on the planet will be exposed to the hazards of solar x-ray and extreme-ultraviolet radiation, bursts of energetic solar particles, and the round-the-clock barrage of extremely energetic cosmic rays.

As noted earlier, a 100 MeV proton can easily pass through an inch and a half of aluminum, and cosmic ray particles can be orders of magnitude more energetic than that. It is therefore likely that the metal skin and frame of the command service modules and lunar landers that were sent on *Apollo* missions to the Moon in the 1960s and 1970s would have provided little if any protection against these most energetic atomic particles.

Galactic cosmic rays, as noted above, pose difficult problems for extended manned missions since they are more energetic and are present all of the time. The greatest challenge is the accumulated dosage of GCRs encountered on long space flights, such as the deep-space missions, two or three years long, that have been envisioned for the manned exploration of Mars. On missions of this duration, the accumulated dosage of galactic cosmic rays would almost surely exceed the threat from solar flares and CMEs. Indeed, cosmic rays may indeed be a show-stopper for manned missions to Mars unless practical ways are found to provide effective shielding against them.

PRINCIPAL HAZARDS OF HUMAN SPACE FLIGHT ON MISSIONS THAT TRAVEL BEYOND THE EARTH'S ATMOSPHERE AND MAGNETOSPHERE

SOURCE	ENERGY PER PARTICLE IN ELECTRON VOLTS	MOST PROBABLE TIME OF OCCURRENCE; AND EXPECTED EXPOSURE	PRINCIPAL CONCERN
SOLAR ENERGETIC PARTICLES from major solar flares and CME-driven shock waves	$10^6 - 10^9$	During periods of high solar activity, which are more likely during or following maximum years of the eleven yr cycle. -------------- 30 minutes to several days for those associated with observed CMEs	Persistent or repeated exposure to unusually intense events
GALACTIC COSMIC RAYS	$10^9 - 10^{20}$	Continuous	Accumulated dosage on space missions lasting longer than a few days or weeks

Current Capabilities

The ultimate goal in studying space weather is an ability to foretell events and conditions on the Sun and in near-Earth space that will produce potentially harmful societal effects, and to do this adequately far in advance and with sufficient accuracy to allow preventive or mitigating actions to be taken. In this sense it is much like the goal of predicting earthquakes or tsunamis, or more appropriately, the daily tasks of today's National Weather Service, but for a region that is 10^{14} times larger: namely, the volume of space that includes the Sun, the Earth, the Moon and ultimately, Mars.

Our ability to attempt forecasts of this kind rests on the foundation of a long history of accomplishments in research, discovery and investigation; on the daily operation of solar and magnetic observatories around the world; on the launch and operation of a growing number of spacecraft designed to meet this need; on the development and continual improvement of analytical models of parts or all of the Sun-Earth system; and on a national and international infrastructure which has helped organize and sustain these global efforts for the common good.

In our own country, the interagency framework which can help meet this challenge has been provided since 1995 by the National Space Weather Program (NSWP) which was organized to ensure collaborative efforts among seven federal agencies that had individually addressed or were significantly affected by space weather. Included are NASA; the National Science Foundation; the

National Oceanic and Atmospheric Administration (NOAA, through the Department of Commerce); the Department of Defense; and the Departments of Energy, Transportation, and the Interior.

Operational Facilities

There are two civilian operational organizations that in this country process and interpret space weather information to provide forecasts and real-time warnings of space weather disturbances for specific users of this information.

The first and more specifically relevant to the needs of human space flight is the Space Radiation Analysis Group (SRAG) operated at the Johnson Space Center in Houston, Texas by NASA's Radiation Health Program: an arm of the Agency that has long tracked and studied the potentially harmful impacts of high-energy, ionizing radiation from the Sun or the Cosmos on human space travelers.

The second addresses impacts of space weather on the broader community of those affected by it, from aircraft operations to communications and electric power systems and geophysical exploration. It is known as the Space Weather Prediction Center, or SWPC, a world solar and space weather monitoring center operated in Boulder, Colorado through the joint efforts of NOAA and the U.S. Air Force.

Each of these two dedicated centers keeps a twenty-four hours a day, seven days a week watch on the Sun and near-Earth space. Each is staffed by trained professionals and function much like Combat Information Centers on ships of war, or the Situation Rooms that somewhere deep within the White House or the Pentagon track national security happenings as they unfold.

⊙   ⊙   ⊙

Is more than one operational space weather patrol necessary? Are two policemen or two firemen better than one? The SRAG and SWPC serve two quite different communities and focus on two separate and urgent needs, and through essential overlap provide enhanced scrutiny and back-up. Each keeps a round-the-clock watch on the face of the Sun, its surface magnetic fields and its outer atmosphere, tracks the birth and development of individual solar active regions, and monitors conditions in near-Earth space. Working cooperatively, they keep a close eye on solar magnetic regions that are likely to erupt in flares and produce CMEs, and track their progress with these eventualities in mind. In the same way they follow the evolution of active prominences and watch for ominous changes in the inner and outer corona, identify CMEs as they first appear, project their outward progress and for those headed in our direction, forecast their time of arrival and expected impacts.

In order to do this, the SRAG and SWPC receive continuous flows of information from solar and magnetic observatories around the world, following the path of a Sun that never sets, as well as data from dedicated spacecraft that continually watch the inner corona in the light of its short-wavelength emission, and with space-borne white-light coronagraphs, the changing form of the outer corona.

⊙   ⊙   ⊙

The SRAG at the Johnson Space Center is responsible for ensuring that the dosage of ionizing radiation received by astronauts from solar eruptions and galactic cosmic rays remains below established safety limits. To do this the activity provides

- continual monitoring of space weather conditions, including information from instruments on board operational satellites such as GOES, SOHO and ACE;
- comprehensive advance and real-time information on radiological exposure during specific manned missions;
- monitoring equipment carried on board to measure the actual radiation environment inside and outside the spacecraft
- extra-vehicular activity (EVA) planning, support and monitoring;
- advance information and updates to mission planners, flight controllers, flight directors and flight surgeons on expected solar behavior and radiation levels;
- contingency responses to likely space weather events during specific missions;
- pre-flight modeling of expected crew exposure, particularly during EVAs;
- evaluation of radiological safety from exposure to radiation-producing equipment on the spacecraft itself;
- tracking of the daily and cumulative doses of radiation received by each member of the spacecraft crew; and
- an interface with support groups within and outside the agency, including the SWPC.

An appreciation of the level of effort and expertise devoted to these challenges at the SRAG can be gained very quickly by logging on to its web site, http://srag-nt.jsc.nasa.gov/. Included there, as well, is a trove of summarized information on space shuttle missions currently planned or in progress and the current status of the International Space Station.

⊙ ⊙ ⊙

Additional streams of incoming information allow the SWPC in Boulder to track conditions in the approaching solar wind and the flow of particles at the top of the Earth's atmosphere that foretell the occurrence of magnetic storms and auroral displays. Other data tell of the changing state of the magnetosphere and the electric currents that flow within it, and of conditions in the thermosphere and layered ionosphere.

Based on an assimilation and considered interpretation of all these data, and aided by analytical models, a day-by-day forecast is prepared for space weather conditions in the next three days.

Among the many users of these current reports of vital information are the Federal Aviation Agency; the Departments of Defense and Homeland Security; NASA, providing a supplemental flow of information throughout the full duration of crewed space missions; the U.S. Nuclear Regulatory Commission and electric power industry in anticipation of possible power interruptions; those involved in geological surveys and exploration; and the many others—including particularly the operators of satellite communications and navigation systems and the broadcast industry—who depend upon reliable and uninterrupted telecommunications. All of the general projections and the background information on which they are based are made immediately available to the whole wide world on the internet (see, for example, http://www.swpc.noaa.gov/)

⊙ ⊙ ⊙

Nearly every impending solar-driven storm—and all of the largest and most threatening—can be foreseen with the advantages of 24-hour observations of the Sun, the considerable help of early warnings from the distant ACE spacecraft and improved understanding of the chain of events that connect solar disturbances to the Earth.

One of the most taxing challenges facing the SWPC and SRAG is an accurate and precise prediction of the date and time of arrival at the Earth, the Moon or Mars, of the particles and fields that are driven outward from these very distant events. A decade ago specific predictions for the Earth were correct about a third of the time. Today the batting average is about .500, which is about the same as was the case with the prediction of severe meteorological storms—then a more mature field of study—in the 1960s.

Advance alerts and warnings issued from the SWPC were instrumental in mitigating the possible impacts from what became known as the "Great Bastille Day flare" of July 14, 2000: a long-lasting release of magnetic energy from the Sun that was followed by three fast-traveling shock waves and the immediate discharge of a monstrous CME that was directed, like a huge cannon ball, precisely at the Earth. When it hit, the momentum of the CME pushed the Sun-facing side of the magnetosphere so close to the Earth that spacecraft in geosynchronous orbits, normally shielded by the magnetosphere, were left, fully exposed, outside it. The SWPC in Boulder proved its worth again with highly useful predictions of what came to be called "the Halloween Day" storms in late October, 2003: another awesome display of solar eruptions and terrestrial effects that were released by the Sun two years after the peak of the most recent solar cycle.

Even the early warning ACE spacecraft was so showered with particles in this event that it ceased to operate for a time. By the next day the largest magnetic storm in eleven years had perturbed the upper atmosphere all around the Earth, setting off low-latitude displays of the aurora borealis and australis.

Probably the most threatening impacts on the surface of the Earth were the effects of unusually strong induced electric currents on power distribution systems that damaged power transformers at twelve power plants in North America. But in this case alerts that were issued by the SWPC in advance of the geomagnetic storm allowed time for power companies—including nuclear power plants—to prepare for the event. As a result there were neither blackouts nor major financial impacts.

The Heliophysics System Observatory

Some twenty-six operational spacecraft circling both the Earth and the Sun—or on further voyages of discovery, far from home—now explore, patrol and monitor the complex, coupled Sun-Earth system. They are there to identify, understand and ultimately predict the major changes on the Sun and in near-Earth space that affect space weather and human endeavor. To do this they are designed and operated to complement each other, and to work together as an ongoing System Observatory in near-Earth space.

NASA develops this bold system as an evolving entity, responding to practical needs, new findings and the necessity to replace ailing or failed elements. Long lead-times are involved in the initial concept and design, operational plan, funding, construction, testing and launch of any new spacecraft. Thus, in order to meet anticipated needs of the System Observatory in this and future decades there are now more than forty other spacecraft at various stages of development—from initial concept to fully funded and under construction—working their way through this long pipeline.

Ships in this potential fleet, like those now in operation, differ in many ways, including (1) their specific mission, and the instruments they carry to accomplish it; (2) their size, weight, scope, anticipated lifetime and expected time of delivery; (3) the nature and complexity of their envisioned operation, including their orbit (or orbits where multiple spacecraft are involved) and the proposed scheme of in-flight control and data management; and (4) their method of launch, and often-shared sources of support, within this country or abroad. These parameters all bear on the cost of the proposed mission.

⊙ ⊙ ⊙

Spacecraft of the System Observatory that are currently in operation—or expected to join the fleet by the end of year 2009—are briefly described in the accompanying table, where they are listed in order of launch date and separated by color code into three broad areas of principal emphasis within the Sun-Earth system.

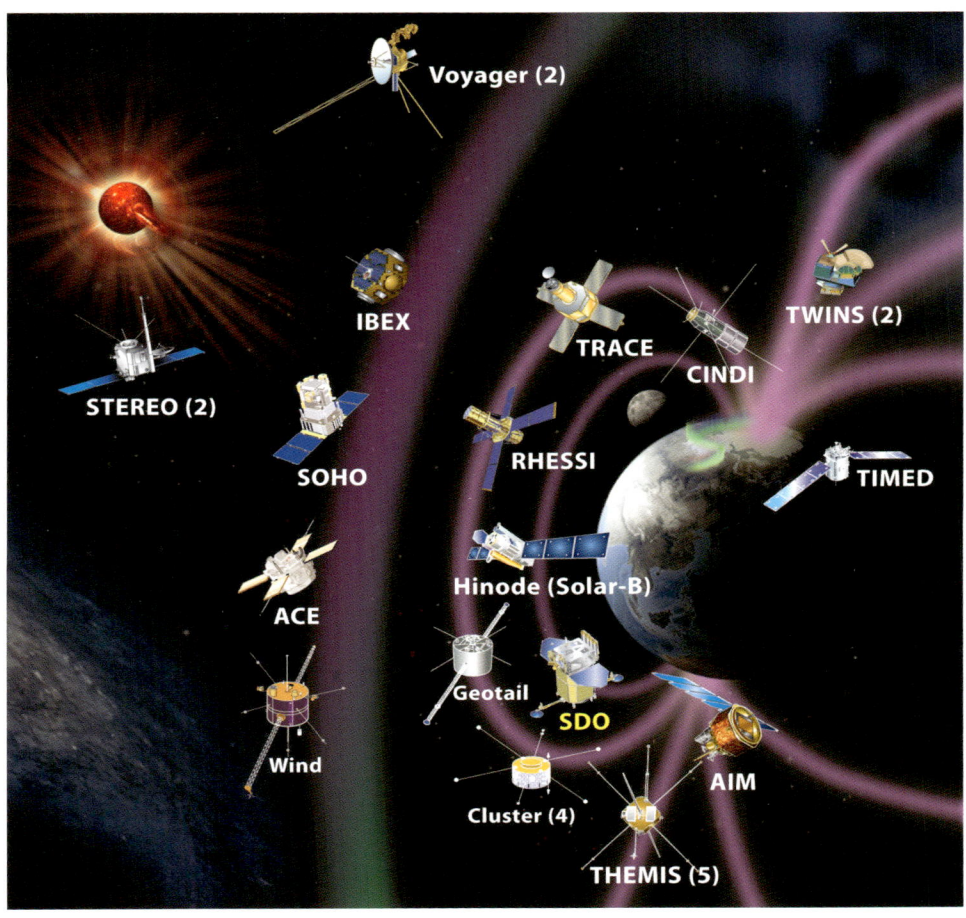

The evolving Heliophysics System Observatory consisting of current and some future spacecraft that study the Sun and the Sun-Earth environment. Currently operating spacecraft missions as of July 2009 appear in white, future scheduled launches in yellow.

SPACECRAFT THAT NOW STUDY AND MONITOR THE SUN-EARTH SYSTEM

Principal emphasis:
Sun and heliosphere
Solar wind particles and fields
Magnetosphere and upper atmosphere

SPACECRAFT	DATE LAUNCHED	ORBIT	MISSION
VOYAGERS 1 and 2 NASA	1977	Outward journey from the Earth and the Sun	Exploration of the outer boundary of the heliosphere
GEOTAIL INSTITUTE OF SPACE AND ASTRONAUTICAL SCIENCE (ISAS) AND NASA	1992	Highly elliptical orbit that takes it from 30 to about 200 R_E	Study of the dynamics of the Earth's magneto-tail
WIND NASA	1994	94,000 miles from the Earth at the L1 Lagrangian point	*In situ* study of the solar wind in a region closer to the Sun than we
SOHO (Solar and Heliospheric Observatory) NASA, ESA	1995	94,000 miles from the Earth at the L1 Lagrangian point	Observations and study of the solar interior, solar irradiance, solar magnetism and CMEs
ACE (Advanced Composition Explorer) NASA	1997	94,000 miles from the Earth at the L1 Lagrangian point	*In situ* measurements of the solar wind before it reaches the Earth
TRACE (Transition Region and Corona Explorer) NASA	1998	Circular, Sun-synchronous orbit around the Earth	Images of magnetic structures in the corona and transition zone
CLUSTER EUROPEAN SPACE AGENCY (ESA) AND NASA	2000	Four spacecraft in elliptical polar orbits, 12,000 to 74,000 miles above the surface	Investigation of critical phenomena in the magnetosphere
TIMED (Thermosphere-Ionosphere-Mesosphere Energetics and Dynamics) NASA	2001	390-mile equatorial circular orbit around the Earth	Study of the coupled magnetosphere, thermosphere and ionosphere
RHESSI (Reuven Ramaty High Energy Solar Spectroscopic Imager) NASA	2002	370-mile Sun-synchronous circular orbit	Study of particle acceleration and energy release in the solar corona and flares

SPACECRAFT THAT NOW STUDY AND MONITOR THE SUN-EARTH SYSTEM (Continued)

Principal emphasis:
Sun and heliosphere
Solar wind particles and fields
Magnetosphere and upper atmosphere

SPACECRAFT	DATE LAUNCHED	ORBIT	MISSION
AIM (Aeronomy of Ice in the Mesosphere) NASA	2007	375 mile Sun-synchronous orbit	Radiometry and images of polar mesospheric clouds; linkages between vertical atmospheric regions
CINDI (Coupled Ion Neutral DYnamics Investigation) U.S. AIR FORCE AND NASA	2008	Equatorial circular orbit within the ionosphere	Measurements of irregularities in the ionospheric plasma as these pertain to radio propagation
HINODE (SOLAR-B) AEROSPACE EXPLORATION AGENCY OF JAPAN (JAXA) AND NASA	2006	370-mile Sun-synchronous circular orbit	Detailed observations and study of the evolution of solar magnetic fields
IBEX (Interstellar Boundary Explorer)	2008	High altitude orbit that reaches 150 thousand miles above the Earth	IBEX images will reveal global properties of the interstellar boundaries that separate our heliosphere from the local interstellar medium
STEREO NASA	2006	Two identical spacecraft in orbit about the Sun at 1 A.U.	3-dimensional observations of the Sun and CMEs
THEMIS (The History of Events and Macroscale Interactions during Substorms) NASA	2007	5 identical spacecraft in equatorial orbits at distances of 10 to 30 R_E	Particles and fields in the tail of the magnetosphere and their connections with magnetic substorms and aurorae
TWINS (Two Wide-angle Imaging Neutral-atom Spectrometers) NASA	2008	Two spacecraft, 29,000 miles above the Earth in high-inclination orbits	Stereoscopic imaging of the magnetosphere to study connections between processes in different regions

SPACECRAFT THAT WILL SOON STUDY AND MONITOR THE SUN-EARTH SYSTEM

SPACECRAFT	LAUNCH DATE	ORBIT	MISSION
SDO (Solar Dynamics Observatory) NASA	Fall 2009	22,200 miles above the Earth in an inclined geosynchronous orbit	Solar observations to clarify the sources of solar variability that affect life and society

From the Earth to the Moon

By Jules Verne

REFLECTIONS

Early in Jules Verne's prescient novel, *From the Earth to the Moon*, published in 1873, the President of the Baltimore Gun Club presents in a formal address to his colleagues his bold plan to fire a manned projectile to the Moon. "*There is no one among you,*" he begins, "*…who has not seen the Moon, or, at least, heard speak of it.*"

And so it is with the even more familiar Sun, on which we so completely depend for the continuance of life on this planet.

Solar Misbehavior

Throughout human history almost all people have seen the Sun as wholly benevolent and indeed the very emblem of dependability and constancy in its daily pattern of rise and set. This longstanding article of faith was to some degree questioned, however, when in 1609 it was first revealed to the western world (what had long been known in the Orient) that when more closely examined the brilliant face of the Sun was not the perfect, unblemished fire most people had always assumed and indeed wanted it to be, but mottled with a scattering of dark spots of many shapes and sizes that came and went from time to time.

This disturbing realization was made in some ways more palatable when more than 200 years later, in 1843—a delay consistent with Tennyson's later reflection that "*Science moves but slowly, slowly; creeping on from point to point*"—it at last came to light that the total number of sunspots seen in any year was not random but strongly periodic, driven by a cycle of about eleven years, which was not unlike the wholly predictable movements of the planets or the comfortable ticking of a clock.

This seemed to imply that if not constant, the Sun was at least regular. But even this qualified consolation was taken away with the later realization that when examined retrospectively over a longer span of time, the Sun behaved in a less ordered and far less predictable way. During the seventy-year period between about 1645 and 1715, for example, sunspot activity almost disappeared altogether, as had happened not long before that between about 1450 and 1540, and in a dozen other instances sprinkled through the last 10,000 years.

In truth, all that we have ever learned about the Sun confirms that though we might wish it otherwise, ours is a variable and moody star. We now know that the total amount of heat and light released into space from its fiery surface varies from day-to-day and year-to-year: rising when the Sun is more active and falling when its surface is less spotted, through a total range of several tenths of a percent. Solar radiation in the invisible and more energetic components of the spectrum, which have the greatest effect on our upper atmosphere, follows a like variation but through far greater excursions, changing in the ultraviolet by several percent and in the far-ultraviolet and x-ray region by factors of 100 to 1000. We also know that isolated regions on the visible surface of the Sun can for a few minutes erupt in an explosive flash of light accompanied by the release into space of a burst of intense short-wave radiation. And that the Sun pours into space and onto the Earth a nonstop flow of energetic atomic particles in the form of a solar wind that blows in all directions throughout the heliosphere; and often throws out whole pieces of itself in the form of colossal ejections of coronal material, some of them headed our way.

⊙   ⊙   ⊙

Solar variability of any kind, gradual or abrupt, alters the amount and nature of the solar energy that reaches the Earth. Moreover, to the best of our knowledge the Sun has been acting this way and doing this to us throughout all of human history, and long before that. Cro-Magnon people could have seen sunspots with their naked eye had they the inclination to look at the dimmed red disk of the Sun when it was near the horizon. Coronal mass ejections repeatedly slammed against the Earth, disrupting its magnetic field and upper atmosphere, throughout the long reigns of the pharaohs; at the time of Christ; and while William Shakespeare was writing plays in London. And although these solar tantrums went long unnoticed, we feel them today and in ever increasing ways.

What Has Changed?

Since prehistoric time what has changed in the age-old Sun-Earth relationship is neither the Sun nor the Earth, but us, through our ever-increasing reliance on technology.

Among the first to sense an immediate effect of sunspots and solar activity—in this case, coronal mass emissions—were early telegraphers in the mid-nineteenth century whose tapped-out dots and dashes were at times garbled in their passage through copper wires that stretched from place to place between wooden poles. The same problem of disruptive electric currents induced by solar-driven magnetic storms was noted at about the same time in the nineteenth century with the use of long communications cables laid down across the sea-floor.

Today, magnetically-induced currents in railroad tracks, in pipelines above and below the surface, and though electrically-conducting soil are a real and growing threat to electric power grids in heavily-populated regions where solar-induced power failures can and have triggered massive blackouts in our interconnected electric power distribution systems, resulting in losses of tens of millions of dollars and more.

Solar-driven changes in the ionosphere, which were previously unrecognized, became a dominant factor in the early 20th century and since, with the advent and ever-wider application of radio waves, affecting early radio broadcasts and other applications of this then emerging technology.

In the world of today solar storms can disturb or interrupt all forms of electronic communication, including essential military, home security, and navigational transmissions and signals sent to and from spacecraft hundreds to millions of miles away. The lifetimes of spacecraft in near-Earth orbits are directly affected by changes in the upper extent and density of the thermosphere, which expands upward with greater solar short-wave radiation, increasing the density of air and hence the amount the spacecraft is slowed by friction at a given altitude. Spacecraft and space equipment in low-Earth orbits are also affected by proximity or passage through the Earth's radiation belts, which are similarly modulated by solar activity. Spacecraft and/or the equipment they carry can be damaged and even put out of operation by highly energetic solar particles. Today commercial and military flight controllers redirect and change the altitude of the flight paths of aircraft to minimize expected exposure to heavy solar particle radiation.

There can be trouble, as well, right here in River City, for solar disturbances can through their impacts on the upper atmosphere disturb many of our common everyday activities, from telephone calls and satellite television to GPS reception and the electronic transfer of information from ATMs, to credit card transactions in stores and restaurants and gasoline pumps.

Acutely vulnerable to drastic changes in solar moods are those manned spaceflights that carry astronauts beyond the protection provided by the Earth's atmosphere and magnetosphere, where they are directly exposed to intense and even fatal particle radiation from major solar flares. This is particularly the case when these brave venturers leave the partial protection of the spacecraft in the course of EVAs, or while on the exposed surfaces of the Moon or Mars. Long voyages to either of these destinations will in addition expose space travelers to a continuous, cumulative dosage of even more energetic galactic cosmic rays, which are also modulated by solar conditions and capable of passing through the spacecraft as though it weren't there.

To detect and monitor significant solar events a fleet of dedicated spacecraft now circle the Earth at various altitudes, serving as a critical outpost of national security. They are there to keep the half of the Sun that is visible to us under continual surveillance, in white light and at wavelengths invisible on the surface of the Earth; as in situ probes to monitor current conditions in near-Earth space; and to serve as distant scouts of what the Sun is sending our way. As important are two dedicated operational centers, manned day and night, where up-to-the minute information from space and the ground are compiled, analyzed, and translated into forecasts and warnings which are made available on the internet in real time to users around the world.

The Sun and Global Warming

The susceptibility of our weather and climate to solar forcing relates directly to another, highly visible issue of current concern around the world. How large a role has the Sun played in the well-documented global warming of the last 50 years, and to what degree could solar changes alter what is expected from an anticipated doubling of greenhouse gases?

Variations in the amount of solar energy received at the top of the atmosphere can through various paths alter meteorological conditions in the lower atmosphere, including the surface temperature of the planet. The degree to which these changes affect land and sea temperatures depends on their magnitude, their persistence, and the relative effect of *competing climate drivers*.

The thermal inertia of the atmosphere and oceans dampens the effect of short term forcing of almost any kind. And on the time scale of days or weeks or months, the impacts of known solar-driven changes are far outweighed by the combined effect of among others, changes in atmospheric circulation and cloudiness; the introduction of volcanic dust and other particulates; and lingering shifts in the interconnected atmosphere-ocean system such as the El Niño effect. On time scales of decades or longer, when the up-and-down effects of these other short-term climate drivers are removed in the averaging process, persistent changes of even a small amount in solar forcing can be recognized. A consistent response to the 11-year solar cycle is found in surface temperature records for both the land and the ocean when these are averaged over large areas and for longer periods of time.

The magnitude of the Sun's impact on global surface temperature was unknown and probably unknowable before it became possible to make highly-precise measurements of the incident solar radiation from above the Earth's atmosphere. What has come from the now 30 years of these measurements is that the total solar radiation varies through limits of about 0.1%, which corresponds to a cyclic change in surface temperature of about 0.1° F.

Possible longer-term, larger-amplitude changes in the Sun's output of energy have been invoked to explain decade-to-century-long trends of warmer or cooler temperatures in climate records of the past—such as the temporal correspondence of the Maunder Minimum in solar activity with the Little Ice Age. Gradual changes in the level of solar activity persisting for 50 to 100 years are indeed readily apparent in both the 400-year record of sunspots and in the much longer tree-ring record of the production of radiocarbon. But the available 30-year sample of measurements of total solar radiation is insufficiently long to allow any meaningful check on whether known longer-term changes in solar activity have been accompanied by concurrent long-term changes in solar radiation. Or whether the amplitude of possible longer term changes in solar radiation might exceed the limits of what characterizes the 11-year modulation.

Nor can a feeling that there might possibly be longer-term, larger solar impacts compete with hard measurements covering almost half a century of an ever-steeper increase in the amount of carbon dioxide in the atmosphere, or mounting evidence that the timing and nature of terrestrial responses, as in Alaska, for example, fit so well what theory predicts for greenhouse warming as opposed to direct solar heating.

As for solar irradiation, the only established fact is that changes that exceed a few tenths of a percent have yet to be found in the limited 30 years of reliable measurements from space. The impact of changes of so small a magnitude on the mean global temperature of the Earth—which have now been observed—are not in the same league with the far larger effects expected from the well-established increase in atmospheric greenhouse gases.

Thus solar forcing of surface temperature is for now relegated to a secondary, second-order effect in terms of its impact on present trends in surface temperature. Any claim that the accelerated global heating of the past 60 years can be attributed to a changing Sun—thus conveniently absolving ourselves of any guilt in the matter—is clearly wishful thinking.

⊙ ⊙ ⊙

Much of what is now known in each of the two practical issues just addressed—our increasing susceptibility to solar changes and space weather in the modern world, and the degree to which the Sun has contributed to global warming—has come to light in but the last few decades.

It is fair to say that progress made toward societal application and reliable prediction in the past 30 years exceeds in both cases all that we had accomplished toward this end in the last 300.

It is also true that the principal contributor to these most recent breakthroughs has been our new-found ability to observe and monitor the Sun, solar irradiation, CMEs, the solar wind, the magnetosphere and the upper atmosphere from the vantage point of space. Coupled with the foresight to plan and the dedication to put in place and keep in place, one by one, the spacecraft and space-borne instruments specifically dedicated to that essential and continuing task.

ACKNOWLEDGMENT

This book owes much to six scientists in different disciplines who helped me immeasurably by critically reviewing the manuscript at different stages of its production: Mark Giampapa, Joe Giacalone, Jack Gosling, Janet Kozyra, Terry Onsager and Chris St. Cyr, although any errors which remain in the final printed version are solely mine.

I am as much indebted to those in the Heliophysics Division of the National Aeronautics and Space Administration who envisioned a need for a popular-level description of the entire interconnected and ever-changing Sun-Earth System, and granted me the opportunity to write one: Dick Fisher and Lika Guhathakurta. I am obliged as well to Steele Hill for help in identifying images; and especially to Jennifer Rumburg, who worked closely with me on the initial conceptual identity of the book, its content, and with artistry and creativity, on all aspects of layout and production.

I am also grateful to the National Solar Observatory in Tucson and particularly to Priscilla Piano of the NSO administrative staff.

Most of all I thank my wife and my best friend, Barbara, for invaluable advice and counsel, daily encouragement and support, and patience without end.

Jack
John A. Eddy

GLOSSARY OF TECHNICAL TERMS

A

absorption line
a dark line in the spectrum of sunlight produced by the absorption of light of a specific wavelength by an identifiable chemical element, such as hydrogen or iron

active prominences
a broad class of solar prominences of a dynamic nature, including eruptive prominences, radially-structured (as opposed to helical) prominences called sprays, and loop-shaped prominences, all of which are generally associated with flares

active region
a complex area of concentrated magnetic field on the surface of the Sun where sunspots, faculae, plages (bright regions in the chromosphere) and filaments occur together

adaptive optics
optical elements in a telescope that are designed to automatically compensate for the effects of atmospheric turbulence and other disturbances

aerosol
a fine, solid or liquid particle suspended in air, such as a dust mote

alpha particle
the nucleus of a helium atom, made up of two protons and two neutrons

anomalous cosmic ray; ACR
a lower-energy cosmic ray created within the outer heliosphere when neutral atoms from the interstellar medium are ionized by solar short-wave radiation

anthropogenic
of human origin

astronomical unit (A.U.)
the average distance from the Earth to the Sun, about 93 million miles

Technical or semi-technical words introduced in the text are printed in blue *generally where they first appear in the text. They are then defined in alphabetical order in the Glossary.*

atom
the smallest unit of a chemical element (such as helium or carbon) that exhibits the properties characteristic of that element

atomic weight
the sum of the number of protons and neutrons in the nucleus of an atom

aurora
diffuse, glowing light emitted from atoms and molecules in the Earth's upper atmosphere when incoming high energy particles from the Sun or the Earth's magnetosphere collide with them

aurora australis
aurorae that appear in the Earth's southern hemisphere

aurora borealis
aurorae that appear in the Earth's northern hemisphere

auroral oval
a belt of latitude, several hundred miles wide and centered on the north or south magnetic pole, in which most aurorae occur

auroral substorms
particle-induced disturbances in the magnetosphere which are linked to the brightest and most dynamic aurorae

---

B

biosphere
that part of the Earth's surface, atmosphere, and oceans where life can exist

bow shock
the shock wave formed in a plasma stream when it comes upon a barrier such as the Earth's magnetosphere, which deflects, slows and heats the incoming plasma. So-named because of its similarity to the curved "bow wave" which forms in the water ahead of the bow of a moving ship

bulk speed; bulk motion
the average speed of individual particles within a streaming cloud of plasma, such as a solar wind stream

C

carbon cycle
the ongoing exchange of carbon among living things, the atmosphere, the oceans and the Earth's surface

Cêrenkov radiation
electromagnetic radiation emitted by an atomic particle traveling through a medium at a speed that exceeds the speed of light in that medium

chromosphere
a highly-structured layer in the Sun's atmosphere that lies between the cooler photosphere and the much hotter corona

chromospheric network
the pattern of bright interconnected lines in the solar chromosphere that delineate the strong magnetic fields that exist at the boundaries of the closely-packed supergranulation cells

climate sensitivity
the change in the Earth's mean surface temperature expected in response to a 1% change in the total solar radiation received at the top of the Earth's atmosphere

closed field line
a magnetic line of force that connects regions of opposite magnetic polarity, as for example, in magnetically-active areas on the Sun, or between the two magnetic poles of the Earth

CME
see coronal mass ejection

convection
the upward transport of energy carried by the rising motion of hotter material, as in the atmosphere of the Earth or the Sun, or in a kettle heated from below

convection cell
a closed cell of gas or liquid within which hotter material is borne upward by convection, at some point releases its thermal energy, and then sinks back downward to be reheated again

convection zone
the deep, internal layer just beneath the solar surface in which convection is the dominant form of energy transport

corona; solar corona
the outermost atmosphere of the Sun, characterized by very low density and extremely high temperature

coronagraph
an especially-designed telescope employed in space or on the ground to allow observation of the Sun's corona at times other than during a total solar eclipse

coronal hole
a dark region in the inner corona where open magnetic field lines allow coronal electrons to escape the Sun, resulting in a region of partially-depleted density and hence reduced brightness

coronal mass ejections; CMEs
segments of the outer corona that have been expelled from the Sun into interplanetary space in the form of expanding clouds of solar plasma, some as large or larger than the Sun itself

cosmic rays
highly energetic atomic nuclei stripped of most or all of their electrons, that pass at all times and from all directions through intergalactic, interstellar and interplanetary space. The most energetic of these (called galactic cosmic rays) presumably originate in dynamic cosmic phenomena within or beyond the galaxy, such as supernova explosions

cosmic ray shower; cascade
the expanding sequence of secondary cosmic rays initiated by the collision of an incoming cosmic ray with an atom or molecule of air. This initial impact produces daughter particles that then collide, farther down in the atmosphere, to create a continuing series (or cascade) of similar events

cosmogenic nuclide
an isotope such as radiocarbon (^{14}C) which is formed in the middle atmosphere of the Earth when an incoming cosmic ray collides with a neutral atom of air

cosmos
the universe, often taken to indicate all of space

Cretaceous period
a period of Earth history at the end of the Mesozoic era, extending from about 135 to 63 million years before the present

D

D, E, and F regions
horizontal strata in the middle and upper atmosphere of the Earth where free electrons are sufficiently abundant to affect the passage of radio and other electromagnetic radiation through these regions: D region @ 40-60 miles; E region 60-80, and F region about 100 to 1000 miles or more above the surface

daytime aurorae
displays of the aurora borealis or australis which occur in the atmosphere above the daylit hemisphere of the Earth, and because of this go largely unseen

density
the ratio of the mass of an object to its volume

differential rotation
rotation of a fluid or non-rigid body like the Sun in which different depths rotate at different rates, with a resultant slippage between them

disk (of the Sun)
the apparent circular shape of the Sun when seen in the sky, resulting from a two-dimensional projection of a spherical object

distant neutral line
a quasi-permanent site in the far tail of the magnetosphere (at a distance of more than 300 thousand miles from the Earth's surface) where lines of magnetic force emanating from the magnetic northern and southern polar regions of the Earth are brought in contact with each other, creating a closed field line in the form of a long loop rooted at high latitudes in the two hemispheres

drag
friction produced when the path of a spacecraft orbiting the Earth takes it through a denser (lower) part of the atmosphere, resulting in a reduction of speed and hence a further decrease in the altitude at which it operates

dynamo
a device through which mechanical energy (e.g., in an electric generator, or the movement of a conductor through a magnetic field) is changed into electric energy. The solar dynamo works in a similar way by harnessing the mechanical energy of differential rotation to twist polar field magnetic lines within the Sun into toroidal fields (perpendicular to the Sun's axis of rotation.) These in turn give rise to sunspots when they are carried upward by convection to the solar surface

E

eccentricity (of an ellipse)
a measure of the extent to which an orbit, such as that of the Earth around the Sun, departs from a perfect circle

ecliptic plane
the plane defined by the path of the Earth's orbit around the Sun, fixed by the 23° tilt of the Earth's axis of rotation. As such, the ecliptic also defines the apparent path followed by the Sun across the celestial sphere in the course of one year: ascending higher in the sky in summer and lower in winter

electric field
a region of space in which a detectable electric intensity is present at every point

effective temperature
the temperature of a portion of the terrestrial or solar atmosphere defined by comparing the amount of energy it radiates with that emitted from an ideal radiator that is heated and cooled by radiation alone

electromagnetic radiation
radiation consisting of oscillating electric and magnetic fields, including gamma rays, visible light, ultraviolet and infrared radiation, radio waves and microwaves

electromagnetic spectrum
the entire range or electromagnetic radiation, from gamma rays to microwaves

electron
a sub-atomic particle with a single negative charge and a mass of less than 1/1000 that of a proton, the corresponding positively-charged particle

electron volt; eV
the energy required to raise an electron through a potential of one volt

electrostatic discharge; ESD
the release of electric energy from a stationary (or static) charge, such as that built up when walking across a woolen rug or collected on the outer, metallic surface of a spacecraft

el Niño, la Niña
opposite phases of a quasi-periodic worldwide climatic change, persisting for

three or more seasons, which is initiated by either a pronounced warming (or for la Niña, a pronounced cooling) of surface waters in the tropical eastern Pacific Ocean

ENSO (El Niño Southern Oscillation)
the combination of three related climatological phenomena: el Niño, la Niña, and the large-scale, seesaw exchange in sea-level air pressure between areas of the western and the southeastern Pacific known as the Southern Oscillation

equatorial orbit
a spacecraft orbit around the Earth that remains within north temperate, equatorial, and south temperate latitudes

erg
a metric unit of energy equal to the work done by a force of one dyne (the force required to accelerate one gram of mass by one centimeter per second^2, or as the late George Abell once described it, about the energy exerted by a fly doing a pushup

EVA (extra-vehicular activity)
manned spaceflight ventures that take astronauts outside the protective structure of the spacecraft, including surface excursions on the Moon or other planets

exosphere
the region of relatively constant high temperature in the Earth's atmosphere above an altitude of about 600 miles. Here the few neutral atoms and molecules that remain are on their way out of the atmosphere, due to their high thermal velocities and the very low density of the region

extreme-ultraviolet; EUV
high-energy, short-wave electromagnetic radiation between wavelengths of 10 and 200 nanometers (100 to 2000 angstroms)

F

facula (plural, faculae)
irregular patches in either the photosphere or chromosphere that appear brighter than their less disturbed surroundings as a result of the weak, vertical magnetic flux tubes that are concentrated there

filament
a chromospheric prominence seen on the disk of the Sun in the light of a hydrogen absorption line as a sinuous line, darker than its surroundings. The same feature observed at the limb would appear as a bright solar prominence

firn
glacial snow that has been partially compressed by thawing and freezing but not yet converted to glacial ice

flare; solar flare
a sudden and highly-localized increase in the brightness and the energy released from a restricted area on the solar surface which is most often located within a complex solar active region; thought to be provoked by instabilities in magnetic structures that cause opposite field lines to reconnect in a very small volume of material. in some ways like a short circuit between two electrical wires

fossil fuel
a carbon-based fuel (such as coal, oil or natural gas) that is formed in the Earth from plant or animal remains of a much earlier era

frequency
the number of electromagnetic waves that cross a given point per unit time, for example 60 per second for U.S. electric power, 97.3 million for an FM radio broadcast

G

galactic cosmic rays; GCRs
see cosmic rays

galaxy
a large assemblage of stars, nebulae, and interstellar gas and dust, of which the Milky Way, the galaxy in which our Sun is located, is an example

gamma rays
the most energetic form of electromagnetic radiation, of shorter wavelength and higher energy than x-rays, which is the next most energetic

gap, or slot region
a depleted zone that separates the Earth's inner and outer radiation belts—extending from about 12,000 to 16,000 miles above the surface of the Earth—in which protons but few if any electrons are found

geosynchronous (geostationary) orbit
an orbit at a distance from the Earth of about 22,200 miles, at which altitude an orbiting object revolves around the equator of the Earth in synchrony with the Earth's rate of rotation. Satellites that orbit the Earth in this way appear fixed in the sky, always remaining above a fixed geographical region on its surface

geotail; geomagnetic tail; magnetotail
the extension of the magnetosphere formed on the night-side of the Earth as its magnetic field is swept downstream by the pressure of the solar wind. The geotail acts as a giant energy reservoir for the magnetosphere and plays an important role in geomagnetic storms and other dynamic processes

global electric circuit
the cumulative effect of all the thunderstorms that charge the ionosphere to a potential of several hundred keV with respect to the Earth's surface. The difference in electric potential that thunderstorms create drives vertical electric currents downward from the ionosphere to the ground in regions where thunderstorms do not occur, completing the circuit

granule
the top of one of the close-packed, subsurface convective cells seen in the solar photosphere; their all-over pattern, called solar granulation, covers the entire photospheric surface

great circle
the intersection of a plane passing through the center of the Earth with its surface, which defines the shortest distance between two points on the surface of the spherical Earth. A great circle course—the path followed in long-distance air and ship travel—will appear as a curved line when plotted on a navigational chart

ground-induced current; geomagnetically-induced current; GIC
sudden surges of 10s to 100s of amperes of direct current that instantly flow through unintended conductors such as ground wires, railroad tracks, pipelines, or underground wires that happen to connect areas of different electric potential on the surface of the Earth. These differences in ground potential can arise immediately from large solar-driven changes in the magnetosphere and ionosphere

H

half-life
the time required for half of the radioactive nuclei or other unstable particles in a sample of material of any kind to disintegrate, which for radiocarbon (^{14}C) is 5,730 years

heavy elements
chemical elements with atomic numbers greater than 2, which includes all

elements in the periodic table other than (and also heavier than) hydrogen and helium, the two most abundant in the Sun

heliopause
the outer boundary of the heliosphere, at which place the solar wind becomes indistinguishable from the local interstellar medium

heliophysics
the comprehensive study of the Sun and the region of space over which it wields influence, involving all scientific disciplines that study the Sun and the Earth, near-Earth space, the inner planets, and the processes that link these together

helioseismology
the study of the Sun's interior deduced from the observable oscillations of its surface

heliosheath
the region of subsonic flow that stands between the termination shock in the extended solar wind and the outer boundary of the heliosphere

heliosphere
the vast region surrounding the Sun dominated by atomic particles and magnetic fields that are carried away from the Sun by the solar wind

high-speed solar wind; high-speed stream
solar particles and imbedded magnetic fields (plasma) driven out from the outer atmosphere of the Sun at speeds of about 400 to more than 500 miles per second, originating from coronal holes and other open-field regions in the corona

Holocene epoch
the roughly 11,000 year period of time in Earth history between the end of the last Ice Age and the present day

hydrologic cycle
the continual, cyclic exchange of the Earth's supply of water among what is held in the atmosphere, the oceans, lakes, and rivers, and the solid Earth

---

I

infrared; infrared radiation
electromagnetic radiation of wavelength longer than visible light (beyond the

darkest red we can see) but shorter than radio waves, and which produces the sensation of heat

inner radiation belt
the innermost of the two concentric Van Allen belts of trapped atomic particles that surround the Earth in the equatorial region. The inner belt contains more energetic electrons, protons and heavier ions and extends upward from the top of the atmosphere to a height of about 12,000 miles

insolation
electromagnetic radiation from the Sun that falls on the top of the Earth's atmosphere

interplanetary magnetic field; IMF
the extension of the magnetic field of the Sun throughout the heliosphere

ion
a neutral atom that has become electrically charged by the addition or loss of one or more electrons

ionization; ionize
any process by which an atom loses electrons

ionizing radiation
high-energy atomic particles (such as cosmic rays) that are capable of dislodging bound electrons from neutral atoms through the force of collisional impact, and thus ionize them. Not to be confused with electromagnetic radiation

ionosphere
the electrically-conducting region in the upper atmosphere made up of three horizontal layers extending from about 35 to more than 1000 miles above the surface, which are produced by the ionization of neutral atoms of air by short-wave solar radiation (see also D, E, and F regions)

irradiance; solar irradiance
electromagnetic radiation in all wavelengths received from the Sun at the top of the Earth's atmosphere in units of energy per square area

isotope
an atom (for example, oxygen-18) that has the same number of protons but a different number of neutrons in its nucleus, and hence a different mass than the common, non-isotopic form, (in this case, oxygen-16)

K

kinetic energy
energy possessed by a body as a result of its motion, equal to half the product of its mass and the square of its speed

kinetic temperature
a temperature directly related to the average speed of atoms or molecules in a substance such as air

L

Lagrangian point
places in the combined gravitational fields of the Sun, Earth and Moon where a spacecraft or other object will not be pulled toward any of them

limb (of the Sun)
the apparent, circular edge of the Sun as we see it in the sky

luminosity
the rate at which electromagnetic energy is emitted from the Sun

M

magnetic declination; magnetic variation
that part of the difference in degrees between the direction of true North and that indicated on a magnetic compass, due to the changing position of the Earth's magnetic pole, which changes with time and location on the Earth

magnetic deviation
that part of the difference in degrees between the direction of true North and that indicated on a magnetic compass at a specific location, which is due to the local presence of iron

magnetic field
the portion of space near a magnetic body (such as the Sun) or a current-carrying body (such as an electric power line) in which there is a detectable magnetic force at every point in the region

magnetic field lines; magnetic lines of force
imaginary lines (like the arrows used to show wind flow direction on a meteorological chart) that indicate the direction of the magnetic force at any

point in a magnetic field. A compass needle aligns itself along these local lines of force of the Earth's field

magnetic reconnection; magnetic merging
a process through which oppositely-directed. closed magnetic field lines come into contact, sever, and join to form new magnetic field structures. In the process part of the magnetic energy contained in the magnetic field is converted into thermal or kinetic energy

magnetic pole
either of two non-fixed points on the Earth, close to but not coincident with the north and south rotational poles, where the Earth's magnetic field is most intense and where magnetic field lines are most nearly perpendicular to the Earth's surface

magnetic storm; (geomagnetic storm)
a severe but transitory fluctuation in the Earth's magnetic field, evident initially as a sharp decrease in the strength of the horizontal component of the Earth's magnetic field, felt around the world and lasting a few hours, followed by a recovery phase lasting a day to several days. Geomagnetic storms are most often initiated when regions of enhanced solar wind flow compress the steady-state form of the magnetosphere on its Sun-facing side

magnetometer
a classical instrument employed since the 1820s to measure the intensity of the Earth's magnetic field at a given point on the surface of the planet, and in modern form, to record minute-to-minute variations as a function of time

magnetopause
the outer boundary of the Earth's magnetosphere, where the strength of the solar wind magnetic field surpasses that of the Earth. Though highly variable, it is typically 40,000 to 60,000 miles away from the Earth on the Sun-facing side, and much farther away on the down-wind side

magnetosheath
the region of slowed, heated and turbulent solar wind that lies between the bow shock in the incoming solar wind and the Earth's magnetopause

magnetosphere
the region around the Earth occupied by its magnetic field

Maunder Minimum
a distinctive period between about AD 1645 and 1715 when sunspots and solar activity were very much depressed

mesopause
the upper boundary of the mesosphere and lower boundary of the thermosphere, which lies just above it

mesosphere
the upper part of the middle atmosphere of the Earth, extending from about 30 to 53 miles above the surface, in which air temperature falls monotonically from about plus 200 to minus 135° F

Milankovitch effect
the combined effects of three subtle changes in the orbit of the Earth and its axis of rotation, acting over periods of tens of thousands of years, which work together to reapportion in time and reallocate in space the continuous flow of energy that the Sun delivers to the Earth

molecule
atomic nuclei and electrons from one or more different elements that are bound together, such as water (H_2O) or molecular oxygen, O_2

muon
an unstable sub-atomic particle with a lifetime of about 10^{-6} seconds, a mass about 200 times greater than an electron, and a negative electrical charge. Muons are among the "secondary particles" released when cosmic rays impact atoms or molecules of air in the upper and middle atmosphere

N

near-Earth neutral line
a region within the Earth's geomagnetic tail, at a distance of about 96,000 miles from the surface at which, during magnetic storms, magnetic field lines of opposite polarity rooted in the two magnetic poles of the Earth come together and reconnect. This results in a closed field line along which solar and other charged particles in the geotail are redirected back toward the Earth

near-infrared
the portion of the infrared spectrum of solar electromagnetic radiation that lies just beyond the limit of the visible portion, extending between wavelengths of about 700 and 7000 nanometers [7000 angstroms to 7 microns]

near-ultraviolet
the portion of the ultraviolet spectrum of solar electromagnetic radiation that adjoins the visible spectrum, extending between wavelengths of about 200 and 400 nanometers [2000 to 4000 angstroms]

neutral line
one of two regions in the magnetotail where stretched-out open magnetic field lines of opposite polarity, attached at the Earth's north and south magnetic poles, are brought in contact, allowing a return path for captive particles in the magnetotail to be channeled back toward the planet

neutron
a subatomic particle with no charge and of mass approximately equal to that of a proton (which is the other principal constituent of the nucleus of any atom)

neutron monitor
a device installed at high-altitude stations to detect the passage of neutrons resulting from collisions between cosmic rays and atoms or molecules in the Earth's atmosphere, as a measure of the flux of cosmic rays that produce these collisions

North Atlantic Oscillation; NAO
a quasi-regular, back-and-forth oscillation in sea-level air pressure between large-scale regions in the North Atlantic Ocean, which in terms of climatic function resembles the oscillatory ENSO phenomenon of the equatorial Pacific Ocean

northern lights
aurorae seen in the Northern hemisphere

nucleus (of an atom)
the heavy part of an atom, composed mostly of protons and neutrons, about which electrons revolve

O

open field line; open magnetic field line
a magnetic line of force in the magnetic field of either the Sun or the Earth, one end of which is rooted in the photosphere or at the surface of the Earth, and the other drawn away and detached by dynamic forces

outer radiation belt
the outer of the two concentric Van Allen belts of trapped atomic particles that surround the Earth in the equatorial region. The outer belt is separated from the inner belt by a 4000 mile gap, and extends above the surface of the planet from about 16,000 to 24,000 (and at times as far as 36,000) miles. Within it are the lighter and less energetic trapped particles: primarily weaker electrons with energies in the range of 10,000 to about one million electron volts

ozone layer; stratospheric ozone layer
the region in the Earth's middle atmosphere, between altitudes of about 25 and 65 miles above the surface, where almost all atmospheric ozone is found. The remainder is created in the form of air pollution at ground level in the photosphere

---

P

penumbra (of a sunspot)
a less-dark, roughly annular region surrounding the dark, cooler central portion of a sunspot, made up of a radial fine structure of dark and bright filaments

photic zone
that region of the upper ocean into which sunlight penetrates, extending (depending upon the clarity of water and the wavelength of light) to about 650 feet below the surface: greatest in the blue and least in the ultraviolet and infrared

photon
a discrete packet of electromagnetic radiation whose energy is directly proportional to the frequency of radiation and is thus much greater for an ultraviolet photon than one in the infrared

photosphere
the region of the solar atmosphere from which all visible light and heat are radiated into space. The intangible surface we see when we look at the Sun in visible light

photosynthesis
the process by which plant cells containing chlorophyll convert incident sunlight into chemical energy, and thereby synthesize needed cellulose for plant growth from carbon dioxide and water, with the release of oxygen

pion
an unstable sub-atomic particle with a lifetime of about 10^{-8} seconds, a mass about 260 times greater than an electron, and either a neutral or negative charge

plage
a bright region seen in the Sun's photosphere or chromosphere

plasma
an often very hot, electrically-neutral gas composed of an approximately equal number of electrons and protons, capable of carrying an imbedded magnetic field

plasmapause
the outer boundary of the Earth's plasmasphere and inner boundary of the magnetosphere

plasma sheet
the central and densest part of the Earth's magnetotail, consisting of a compressed sheet that extends downwind of the Sun for at least 950,000 miles from the Earth, separating the northern and southern lobes of the tail, which have opposite magnetic polarity. The plasma sheet is a major storage region for ionized particles in the geomagnetic tail

plasmasphere
the upward extension of the Earth's ionosphere into the exosphere, within and co-existing with the magnetosphere, which reaches on its Sun-facing side from about 1000 miles to as much as a million miles above the surface. It consists of a relatively low-energy plasma and takes its form as charged particles from the ionosphere flow upward to fill the relative vacuum of space surrounding the Earth

plate tectonics
the movement of segments, or plates, of the encrusted, outer layers of the solid Earth over the hotter and more fluid underlying mantle

Pleistocene epoch
the earlier of the two epochs of the Quaternary period of Earth history, extending from about a million years until 11,000 years before the present and characterized by the alternate appearance and recession of northern glaciation

polar cap
in auroral nomenclature, the area around either the north or south magnetic pole of the Earth bounded by the inner boundary of the auroral oval. Here, poleward of the auroral oval, auroras are more frequent but weaker, more diffuse and less variable than in the auroral oval itself

polar cap aurora
aurorae, seen throughout the solar cycle, that occur outside the bounded region of the auroral oval, in the space that separates it from the magnetic pole

polar cusps
the singular regions over the Earth's magnetic poles where magnetic field lines are nearly perpendicular to the Earth's surface, creating an "opening" in the magnetosphere that allows charged particles paths of easier entry into the upper atmosphere

polar orbit (around the Earth)
a spacecraft orbit that passes over the polar (as opposed to restricted to middle and equatorial) latitudes of the Earth. For spaceborne instruments that observe the Sun, a polar orbit offers the opportunity for continuous 24-hour monitoring. The present polar orbit of the Ulysses spacecraft around the Sun was chosen to allow the first observations of much of the solar poles and to explore conditions in the solar wind above them

polar plumes
coronal rays extending like a crown outward from the two magnetic poles of the Sun, that outline the open magnetic field lines which emanate there

positive ion
an ion formed by the loss of one or more of the electrons from a neutral atom, resulting in an atomic particle of net positive charge

primary cosmic ray
a high-energy atomic particle that arrives at the Earth from beyond the planet, as opposed to the secondary cosmic rays that are formed as a result of a collision of a primary with an atom or molecule of air in the Earth's atmosphere

prominence; solar prominence
a shapely extension of chromospheric material into the corona, formed and suspended there by magnetic fields, which is visible at the limb of the Sun or on the disk (where it is called a solar filament)

proton
heavy subatomic particle of unit positive charge which is one of the two principal constituents (the other, neutrons) of the nuclei of atoms

proxy record
data regarding past solar behavior that is obtained indirectly, derived from what is known of the Sun's impacts on other physical phenomena such as aurorae or records of tree-ring radiocarbon or ice-core beryllium-10

Q

quiescent prominence
a long-lived and relatively stationary solar prominence, in contrast to those that are more active or eruptive

R

radiant energy
electromagnetic energy emitted from the Sun

radiation
here used to mean electromagnetic energy, as defined above. When qualified as "particle radiation" it is often conventionally used to refer to the kinetic energy of high energy particles

radiation belts; Van Allen belts
two concentric areas of trapped electrons, protons and ions held within the closed part of the Earth's magnetic field, from about 600 miles to 25,000 miles above the surface

radiative zone
a voluminous region in the solar interior, between the innermost core and the outer convective zone, where radiation is the dominant mode of outward energy transport

radio frequency spectrum
the lowest frequency, lowest energy portion of the total spectrum of electromagnetic radiation from the Sun, which includes radio waves of all wavelengths

radio telescope
an antenna or set of antennas designed to make observations of the Sun or other astronomical objects based on their emission of radio waves

radio waves
electromagnetic radiation emitted in the radio frequency spectrum

red giant
a star that is larger and cooler than the Sun

resolution
spatial resolution: the degree to which a telescope can distinguish, or resolve, fine details in the image it produces, expressed in terms of an angle in the sky, such as 1 arc second; temporal resolution: the shortest interval of time for which information can be distinguished

ring current
an electrical current produced in the equatorial plane within the closed part of the Earth's magnetic field where properties of the magnetic field cause ions and electrons to drift in opposite directions

rotation
the turning of the Earth (one turn in 24 hours) or the Sun (in about 27 days) about an axis that passes through it

S

scattering
the dispersal of a beam of light into a spread of directions as a result of physical interactions: in the daytime sky, the redirection of incoming sunlight across the dome of the sky by its interaction with molecules of air; in the white-light corona, the redirection of photospheric radiation by free electrons

scintillation
a flickering of electromagnetic radiation caused by its passage through turbulent media: examples are the twinkling of stars caused by the passage of starlight through turbulence and inhomogeneities in the air; and irregular deflections in the passage of radio waves through layers of electrons in the ionosphere

secondary cosmic ray
a secondary or "daughter" particle produced by collisions between primary cosmic rays from space and the atomic nuclei of atoms and molecules in the Earth's atmosphere

sectors
discrete, wedge-shaped segments, centered on the Sun, in the expanding solar wind in which the magnetic polarity of the source region on the Sun is carried outward in the plasma and preserved. They are sensed at the Earth (as they sweep by with solar rotation) as distinct changes in the prevailing polarity of the solar wind

seismology
the study of earthquakes and the internal structure of the solid Earth deduced from the analysis of reflected sound waves

shock wave
an abrupt change in temperature, speed, density and pressure in a moving plasma, produced by the movement of an object traveling through the medium at a speed greater than the local speed of sound, that can accelerate energetic particles and trigger geomagnetic phenomena. A similar phenomenon occurs in the Earth's atmosphere when an aircraft reaches supersonic speed

short-wave radiation
electromagnetic radiation from the Sun in the ultraviolet, x-ray and gamma ray region of the spectrum

single event effect; SEE
a malfunction or failure of a piece of electronic equipment in a spacecraft that is traceable to the impact of a single high-energy particle such as a solar energetic proton or cosmic ray particle

single event upset; SEU
the malfunction or failure in a micro-circuit most often traced to the action of a single incoming heavy ion, from the Sun or the cosmos that deposits sufficient electrical charge on a sensitive circuit element to cause it to change state

slow-speed solar wind; slow speed stream
solar particles and imbedded magnetic fields (plasma) driven outward from the outer atmosphere of the Sun at an average speed of about 200 miles per second, most often from coronal streamers

solar activity
phenomena on the Sun such as sunspots, plages, flares, and CMEs whose frequency of occurrence is related to the 11-year sunspot cycle

solar atmosphere
the photosphere, chromosphere, and corona of the Sun

solar constant
the total amount of radiant energy received from the Sun per unit time per unit area at the top of the Earth's atmosphere, at mean Sun-Earth distance. Once thought, erroneously, to be constant, the term has now been supplanted by the more precise term, total solar irradiance

solar cycle; sunspot cycle
the roughly 11-year cyclic variation in the state of activity on the Sun, most apparent in annual averages of the number of sunspots seen on its white-light surface

solar dynamo
the internal mechanism that generates sunspots and magnetically active regions on the surface of the Sun through the interaction of convection, differential rotation, and magnetic fields within the solar interior

solar energetic proton; SEP
a proton ejected from the Sun with an energy in the 1 to 500 meV range, which is potentially damaging due to its heavy mass and high speed

solar equator
a great circle on the Sun, midway between its rotational poles

solar granulation
the pattern of closely-packed convective cells that covers the photospheric surface of the Sun

solar interior
the bulk of the Sun that lies beneath the photosphere, consisting of its nuclear core and overlying radiation and convective zones

solar irradiance
radiant energy received from the Sun per unit time and unit area at the top of the Earth's atmosphere

solar magnetic cycle
the 22-year combination of two eleven-year cycles of solar activity needed for the surface magnetic field of the Sun to switch from one polarity to the other and back again

solar system
the Sun together with the planets and all other objects that revolve about it

solar wind
the continual release of atomic particles and imbedded magnetic fields (plasma) from the Sun resulting from the thermal expansion of the corona

South Atlantic Anomaly; SAA
a region located over southern South America and the South Atlantic Ocean where the strength of the Earth's magnetic field is considerably reduced. Because of this, energetic particles in the Earth's inner radiation belt are able to penetrate here more deeply into the thermosphere, to altitudes where spacecraft operate

space weather
the variable state of the magnetosphere, ionosphere and near-Earth space as perturbed by solar activity and the solar wind: the counterpart of meteorological weather

space climate
long-term or average conditions in the magnetosphere, ionosphere and near-Earth space: the counterpart of terrestrial climate

spectral irradiance; solar spectral irradiance
electromagnetic radiation in specific wavelengths received from the Sun at the top of the Earth's atmosphere

spectrum
the distribution of electromagnetic energy emitted by a radiant source such as

the Sun, arranged in order of wavelength, from gamma rays and x-rays to long-wave radio emission

spicule
one of many spike-like jets of rising gas in the chromosphere, about 600 miles wide and about ten times as high, with a lifetime of about 15 minutes

Spörer Minimum
a distinctive period between about AD 1450 and 1540. evident in the record of tree-ring radiocarbon, when the level of solar activity was much reduced

stratopause
the upper limit of the Earth's stratosphere and lower limit of the mesosphere

stratosphere
the atmosphere extending above the troposphere to an altitude of about 30 miles that exhibits warming with height, the result of the absorption of solar radiant energy by stratospheric ozone

streamer; coronal streamer
a major structure in the outer corona consisting of a magnetically-formed bulbous base near the Sun that is reduced in diameter and swept outward by the solar wind into an extended tapered shape

sunspot
a distinctive, activity-related region in the photosphere, the embodiment of a very strong magnetic field that is cooler and hence darker than the surrounding photosphere

sunspot cycle; solar cycle
the roughly 11-year cycle of variation in the number of sunspots visible on the solar disk: a manifestation of a more fundamental cyclic variation in the number of solar magnetic fields that are brought to the surface from the interior of the Sun

sunspot number
an historical index of solar activity defined in the 1860s as the number of spots that are visible on the Sun at any time plus ten times the number of groups of sunspots, multiplied by a factor intended to correct for differences in telescopes, observing sites, and observers

sunspot maximum or minimum; solar maximum or minimum
the years when the 11-year sunspot cycle reaches its maximum or minimum level

supergiant
a very large, extremely luminous star

super-granulation
convective cells-each of which is up to 200 times larger than those found in the photosphere—which covers the entire surface of the chromosphere in an all-over closely-packed pattern. Like the granules seen immediately below them in the photosphere, they are a consequence of the tumultuous release of energy from the interior convective zone of the Sun, and the organizing of large numbers of photospheric cells into a far larger pattern

suprathermal particles
highly accelerated atomic particles with energies of up to a million or more electron volts, corresponding to temperatures of billions of degrees

T

tachocline
the shear layer in the solar interior between the inner radiation zone and the convective zone that lies above it, thought to be the site of origin of the solar dynamo

termination shock
a shock wave that forms at the place in the outer heliosphere where the solar wind first begins to feel the competing force of stellar winds. In passing through it, the solar wind slows from supersonic (about a million miles per hour) to subsonic speeds

terrestrial
pertaining to the Earth and its inhabitants

Tertiary period
the period of Earth history following the Cretaceous and preceding the Quaternary, extending from about 63 million years until about 2 million years before the present

thermal electrons
electrons with energies per particle in the range from a few electron volts (eV) to 100, representing temperatures of 10,000 to $10^{6}°$ F, typical of particles found in the chromosphere, transition zone, and corona of the Sun

thermal energy
energy associated with the temperature-driven movements of atoms or molecules in a substance

thermal inertia
a property of matter expressing its capacity to retain heat

thermohaline circulation
vertical ocean circulation driven by differences in the temperature and salinity of seawater

thermosphere
the uppermost layer of the Earth's atmosphere, extending from an altitude of about 50 to more than 1000 miles, where absorption of short-wave solar radiation heats the gas to very high temperatures

toroidal; toroid
a surface generated by the rotation of a closed curve, such as a circle, about an axis lying in its own plane, creating a three-dimensional object like a doughnut

total solar eclipse; total eclipse
an eclipse of the Sun in which the photosphere is entirely covered by the lunar disk, darkening the sky and allowing observers within the moving shadow of the Moon to see the dim solar corona

total solar irradiance
electromagnetic energy in all wavelengths received from the Sun at the top of the Earth's atmosphere

transition zone
the thin shell between the Sun's chromosphere and corona, where the temperature climbs from about 10,000 to more than a million° F

Trojan asteroid
one of a group of asteroids (minor planets) that share Jupiter's orbit about the Sun

tropopause
the upper limit of the troposphere and lower limit of the stratosphere, at an altitude of about seven miles above sea-level, though somewhat higher in the tropics

troposphere
the lowest region of the Earth's atmosphere, extending from the surface to the tropopause (about 7 miles high) and characterized by decreasing temperature with increasing altitude; the locus of all weather and climate

U

ultraviolet radiation
invisible, electromagnetic radiation of shorter wavelength and greater energy per photon than that of visible light, spanning wavelengths between about 100 to 4000 angstroms (10 to 400 nanometers)

ultraviolet spectrum
that portion of the spectrum of electromagnetic radiation that extends from about 100 to 4000 angstroms

umbra (of a sunspot)
the dark central region of a sunspot that is often circumscribed by a slightly more luminous penumbra. Sunspot umbrae define intrusions in the photosphere of highly-concentrated, local magnetic fields that inhibit the flow of energy from below and are hence cooler and less bright than the surrounding photosphere

universe
all space, including the totality of all matter and radiation

UV-A and UV-B
two adjoining bands of potentially-damaging incoming solar ultraviolet radiation that can reach the surface of the Earth, as opposed to the shorter wavelength and more energetic UV-C which is entirely blocked by atmospheric oxygen high in the atmosphere. Direct exposure to either UV-A (3150 to 4000 angstroms wavelength) or UV-B (2800 to 3150 angstroms) radiation can damage the skin and eyes and human immune system, although the more energetic UV-B is the greater threat in terms of potential skin cancer

V

visible spectrum
the portion of the spectrum of electromagnetic radiation that can be sensed by the human eye, including all colors of the rainbow from the barely visible violet (about 3900 angstroms wavelength) to the dark and barely visible red (about 6600 angstroms)

W

wavelength
a metric used to distinguish different parts of the spectrum of electromagnetic

radiation, equal to the distance separating two successive crests in a wave of radiation

UNITS OF WAVELENGTH

UNIT	IN METERS	EQUIVALENT	REGION OF ELECTROMAGNETIC SPECTRUM
meter (m)		100 cm	radio waves
centimeter (cm)	10^{-2} meters	10 mm	radio waves
millimeter (mm)	10^{-3} meters	1000 μ	microwaves
micron (μ)	10^{-6} meters	1000 nm	infrared waves
nanometer (nm)	10^{-9} meters	10 Å	visible and ultraviolet waves
angstrom (Å)	10^{-10} meters		visible, ultraviolet, x-ray and γ-rays

wave-particle interaction
an interaction between a charged atomic particle and ambient electromagnetic radiation that can alter the energy of the particle or disturb its path and manner of movement. This can occur in the presence of electromagnetic radiation of a frequency that happens to resonate with the particle's own motion

white-light
the combination of light of all colors in the visible spectrum of electromagnetic radiation. The disk of the Sun, which appears white to us, is an example, as is the white color of clouds or the solar corona, both of which represent scattered light from the white photosphere

X

x-rays
extremely high energy, short-wave electromagnetic radiation in the wavelength range from 1 to 100 angstroms

SOURCES FOR ADDITIONAL INFORMATION

Books

Nearest Star: The Surprising Science of our Sun
by L. Golub and J. Pasachoff; Harvard University Press, Cambridge, MA, 2001, 267 pp. A popular-level book about the Sun written by two renowned solar scientists.

The Cambridge Encyclopedia of the Sun
by K.R. Lang; Cambridge University Press, Cambridge, England, 2001, 256 pp. A complete, modern guide to our nearest star, well-illustrated and explained, in a handsome, over-sized book.

Storms from the Sun: the Emerging Science of Space Weather
by M. Carlowicz and R. Lopez; Joseph Henry Press, Washington, D.C., 2002, 234 pp. A description of the active Sun and space weather and their effects on the Earth, written at a popular level with emphasis on specific effects of dynamic solar events on the Earth and society.

Space Weather
edited by P.L. Song, H.J. Singer, and G. L. Siscoe; American Geophysical Union, Washington, D.C., 440 pp, 2001 A collection of technical articles dealing with many aspects of space weather.

Effects of Space Weather on Technology Infrastructure
edited by I. A. Daglis; Kluwer Academic Publications, Dordrecht, the Netherlands. 334 pp. 2004. Review articles dealing with various aspects of the impact of space weather on human activities, from commercial aircraft to electric power systems, from a NATO workshop on this subject.

Articles

The Sun: Living with a Stormy Star
by Curt Suplee; in National Geographic, July 2004, pp. 2-33. A current portrayal of the Sun, solar activity and space weather, illustrated as only the Geographic can with an awesome collection of diagrams and breath-taking pictures from space.

Living with a Variable Sun
by Judith Lean; in Physics Today, June, 2005, pp. 32-38. An easily-read and up-to-date review of solar variability and its effects on the Earth, told by an acknowledged expert in the field.

Shielding Space Travelers
by Eugene Parker; in Scientific American, March, 2006, pp. 40-47. A popular-level review of the problem of galactic cosmic rays in space travel, by the scientist who accurately foretold the existence and characteristics of the solar wind before it was first discovered.

Some Recommended Web Sites

http://heliophysics.nasa.gov/
A general description of NASA's program that keeps a round-the-clock watch on the Sun and its impacts on the Earth and its near environment, including a summary of current Sun-Earth spacecraft, planned or underway, with access to specific information about each of them.

http://www.swpc.noaa.gov/
Up-to-the-minute information from NOAA's 24-hour Space Weather Prediction Center, including current information on solar activity and conditions in near-Earth space, space weather alerts and warnings that are disseminated worldwide, and current and predicted sunspot numbers for 11-year solar cycle #23, now ending, and #24, soon to be underway.

http://srag-nt.jsc.nasa.gov/
Information regarding the monitoring of space-weather at the NASA Space Radiation Analysis Group at the Johnson Space Center, tailored to the specific needs of individual manned space flight missions, including a trove of summarized information on space shuttle missions currently planned or in progress and the current status of the International Space Station.

http://umbra.nascom.nasa.gov/images/latest.html
Current, daily images of the solar disk, in color and enlargeable to full screen-size portraying today's photosphere, chromosphere, transition region, corona, magnetic fields and vertical motions derived from Doppler images.

http://www.scostep.ucar.edu
A series of delightfully drawn, up-to-date and expertly written introductory explanations of different features of the Sun-Earth system, presented in comic book form, each about 12 pages long. Ingeniously produced in Japan, in English and other languages, and sponsored by SCOSTEP, the International Scientific Committee on Solar-Terrestrial Physics. The most recent issue treats the subject of solar variability and climate. Try them; you'll like them. size;

TABLES

The Planets as Seen from the Sun	15
Solar Temperatures	37
Solar Wind Velocities	50
Manifestations of Solar Activity	69
Earth's Radiation Belts	90
Solar Energy Received at the Earth	103
Energy and Corresponding Temperature of Atomic Particles	112
Solar Particles that Impinge upon the Earth	113
Distances in Units of the Radius of the Earth	129
Price of Admission for Charged Atomic Particles	132
Numbers of Solar and Cosmic Particles Incident Upon the Earth	135
Impacts of Solar Variability and Their Origin on the Sun	140
Energies of Charged Atomic Particles that Strike the Earth	142
Three Stages of a Magnetic Storm	146
Types of Auroral Displays	153
Upper Thermosphere Temperatures	154
Layers of the Ionosphere	159
Origins of Solar Radiation and Regions Affected	160
Impacts of Solar Variability and Their Societal Effects	162
Dose Rate of Galactic Cosmic Radiation at Aircraft Altitudes	169
Aircraft Altitudes in Relation to Atmospheric Features	170
Flight Paths and Hazards of Aircraft and Spacecraft	175
Manned Space Flights which have left the Magnetosphere	178
Orbital Altitudes and Proximate Atmospheric Features	187
Some Major Electric Power Disruptions	203
Recent Prolonged Episodes of Unusual Levels of Solar Activity	216
Useful Space Weather Predictions	238
Travel Times from the Sun to the Earth	241
Principal Hazards of Traveling Beyond the Magnetosphere	245
Spacecraft that now Study and Monitor the Sun-Earth System	252, 253

IMAGES AND ILLUSTRATIONS

Page No.	Source	Page No.	Source
viii	Hubble Space Telescope	88	NASA
2	L. Kangas	98	NASA
12	SOHO, NASA/ESA	101	NASA from J. Lean
18	NASA	110	SOHO, NASA/ESA
21	J. Eddy	120	Leonard Burlaga
22	J. Eddy	121	NASA
25	SOHO, NASA/ESA	125	NASA from S.A. Fuselier
27	J. Eddy	134	J. Bieber
28	Swedish Solar Telescope	138	Jan Curtis
33	SOHO, NASA/ESA	144	Y. Kamide
39	HAO	147	Jan Curtis
41	J. Eddy	149	Polar Spacecraft, NASA
42	NASA	152	NASA
44	NASA	164	NASA
48	SOHO, TRACE, NASA/ESA	169	Captain J.B.L. Jones
49	SOHO, NASA/ESA	179	NASA
52-53	NASA	208	Michel Tourney
54	SOHO, NASA/ESA	210	University of Colorado Library
55	YOHKOH, ISAS	211	Astronomical Society of the Pacific
55	NASA		
59	SOHO, NASA/ESA	214	J. Lean
61	SOHO, NASA/ESA	218	J. Lean
62	SOHO, NASA/ESA	220	L.L. Hood
63	TRACE	222	Douglass & Clader
67	SMM, NASA	226	J. Lean
68	SOHO, NASA/ESA	231	G. Bond, et. al.
70	NASA	234	NASA
73	NASA	244	NASA
75	NASA	251	NASA
83	National Academies Press	254	J. Eddy

INDEX

absorption line	221
active prominence	64
active region	50
adaptive optics	26
Advanced Technology Solar Telescope	26
aerosol	105
alpha particle	132
anamolous cosmic ray; ACR	131
anthropogenic	222
astronomical unit (A.U.)	253
atom	1
atomic weight	188
aurora	7
aurora australis	6
aurora borealis	6
auroral oval	151
auroral substorms	148
biosphere	159
bow shock	85
bulk speed	118
bulk motion	115
carbon cycle	159
Cassini spacecraft	42
Cêrenkov radiation	174
chromosphere	34
chromospheric network	36
climate sensitivity	221
closed field line	50
CME	40
convection	53
convection cell	13
convection zone	31
Copernicus, Nicholas	22
corona; solar corona	35
coronagraph	40
coronal hole	118
coronal mass ejection; CME	40
coronal streamers	36
cosmic rays	1
cosmic ray shower, cascade	133
cosmogenic nuclide	225
cosmos	1
Cretaceous period	58
D, E, and F-regions	157
daytime aurorae	152
density	4
differential rotation	32
disk (of the Sun)	20
distant neutral line	127
dynamo	33
eccentricity (of an ellipse)	106
electric field	145
effective temperature	74
electromagnetic radiation	71
electromagnetic spectrum	60
electron	1
electron volt; EV	87
electrostatic discharge; ESD	192
elliptical	106
El Niño, la Niña	160
ENSO (El Niño Southern Oscillation)	224
equatorial orbit	184
eruptive prominences	25
EVA (extra-vehicular activity)	166
exosphere	81
extreme ultraviolet; EUV	102
facula (plural, faculae)	23
filament	60
firn	229
flare; solar flare	57
fossil fuel	3
frequency	195
galactic cosmic rays, GCRs	131
galaxy	1
Galileo Galilei	20
gamma rays	4
gap, or slot region	89
geomagnetic storm (see magnetic storm)	7
geosynchronous (geostationary) orbit	94
geotail; geomagnetic tail; magnetotail	124
global electric circuit	155
Goldsmid, Johann	20
granule	30
great circle	169
ground-induced current; geomagnetically induced current, GIC	201
half-life	227
Harriot, Thomas	20
heavy elements	83
heliopause	92
heliophysics	9

299

helioseismology	31
heliosheath	96
heliosphere	45
high-speed solar wind; high-speed stream	148
Holocene epoch	230
hydrologic cycle	159
infrared; infrared radiation	19
inner radiation belt	89
insolation	225
interplanetary magnetic field; IMF	143
ion	4
ionization; ionize	35
ionizing radiation	168
ionosphere	79
isotope	219
kinetic energy	47
kinetic temperature	74
Lagrangian point	185
limb (of the Sun)	29
luminosity	18
magnetic field	4
magnetic field lines; magnetic lines of force	31
magnetic reconnection; magnetic merging	47
magnetic pole	7
magnetic storm (geomagnetic storm)	7
magnetometer	205
magnetopause	82
magnetosheath	124
magnetosphere	71
Maunder Minimum	56
mesopause	78
mesosphere	78
Milankovitch effect	107
molecule	1
Muir, John	10
muon	133
NASA, est. 1958	6
near-Earth neutral line	127
near-infrared	19
near-ultraviolet	77
neutral line	126
neutron	45
neutron monitor	133
North Atlantic Oscillation; NAO	224
northern lights	148
nucleus (of an atom)	182
open field line; open magnetic field line	50
outer radiation belt	89
ozone layer; stratospheric ozone layer	74

size; penumbra (of a sunspot)	27
photic zone	225
photon	102
photosphere	13
photosynthesis	3
pion	133
plage	69
planets, seen from Sun	15
plasma	35
plasmapause	92
plasma sheet	124
plasmasphere	91
plate tectonics	159
Pleistocene epoch	131
polar cap	126
polar cap aurora	152
polar cusps	86
polar orbit (around the Earth)	94
polar plumes	37
positive ion	87
primary cosmic ray	90
prominence; solar prominence	36
proton	1
proxy record; proxy data	56
quiescent prominence	64
radiant energy	17
radiation	15
radiation belts; Van Allen belts	88
radiative zone	32
radio frequency spectrum	103
radio telescope	239
radio waves	1
red giant	16
resolution	40
ring current	145
rotation (of the Earth)	173
satellite drag	184
Saturn	21
Scheiner, Christopher	20
scattering	21
scintillation	197
secondary cosmic ray	90
sectors	46
seismology	31
shock wave	113
short-wave radiation	4
single event effect; SEE	192
single event upset; SEU	192
skylight, color of	24
slow-speed solar wind; slow speed stream	49
solar activity	51
solar atmosphere	34
solar constant	99

solar cycle; sunspot cycle	5
solar dynamo	33
solar energetic proton; SEP	115
solar equator	13
solar granulation	30
solar interior	14
solar irradiance	212
solar magnetic cycle	52
solar system	1
solar wind	45
South Atlantic Anomoly; SAA	174
space weather	235
space climate	235
spectrum	17
spicule	36
Spörer Minimum	56
SRAG, Space Radiation Analysis Group	246
stratopause	77
stratosphere	76
streamer; coronal streamer	36
Sun, age of	15
Sun, energy source	3
Sun, expected lifetime	16
sunshine tax	15
Sun, power in kilowatts	15
sunspot	14
sunspot cycle; solar cycle	19
sunspot number	51
sunspot maximum or minimum; solar maximum or minimum	5
Sun, temperatures of	16
supergiant	13
super-granulation cells	36
suprathermal particles	91
SWPC, Space Weather Prediction Center	246
Swedish Solar Telescope	28
System Observatory	9
tachocline	33
termination shock	93
Tertiary period	58
thermal electrons	111
thermal energy	113
thermal inertia	79
thermohaline circulation	224
thermosphere	78
Thompson, Francis	10
toroidal, toroid	32
total solar eclipse; total eclipse	34
total solar irradiance	99
transition zone	35
Trojan asteroid	186
tropopause	76
troposphere	75
ultraviolet radiation	172
ultraviolet spectrum	17
umbra (of a sunspot)	27
universe	15
UV-A and UV-B	6
Van Allen radiation belts	88
visible spectrum	17
wavelength	17
wave-particle interaction	89
white light	24
x-rays	172

The Index is intended as a guide for locating the initial or principal reference for specific technical words or phrases.